IFCoLog Journal of Logics and their Applications

Volume 4, Number 9

October 2017

Disclaimer

Statements of fact and opinion in the articles in IfCoLog Journal of Logics and their Applications are those of the respective authors and contributors and not of the IfCoLog Journal of Logics and their Applications or of College Publications. Neither College Publications nor the IfCoLog Journal of Logics and their Applications make any representation, express or implied, in respect of the accuracy of the material in this journal and cannot accept any legal responsibility or liability for any errors or omissions that may be made. The reader should make his/her own evaluation as to the appropriateness or otherwise of any experimental technique described.

© Individual authors and College Publications 2017
All rights reserved.

ISBN 978-1-84890-265-7
ISSN (E) 2055-3714
ISSN (P) 2055-3706

College Publications
Scientific Director: Dov Gabbay
Managing Director: Jane Spurr

http://www.collegepublications.co.uk

Printed by Lightning Source, Milton Keynes, UK

All rights reserved. No part of this publication may be reproduced, stored in a retrieval system or transmitted in any form, or by any means, electronic, mechanical, photocopying, recording or otherwise without prior permission, in writing, from the publisher.

Editorial Board

Editors-in-Chief
Dov M. Gabbay and Jörg Siekmann

Marcello D'Agostino
Natasha Alechina
Sandra Alves
Arnon Avron
Jan Broersen
Martin Caminada
Balder ten Cate
Agata Ciabttoni
Robin Cooper
Luis Farinas del Cerro
Esther David
Didier Dubois
PM Dung
Amy Felty
David Fernandez Duque
Jan van Eijck

Melvin Fitting
Michael Gabbay
Murdoch Gabbay
Thomas F. Gordon
Wesley H. Holliday
Sara Kalvala
Shalom Lappin
Beishui Liao
David Makinson
George Metcalfe
Claudia Nalon
Valeria de Paiva
Jeff Paris
David Pearce
Brigitte Pientka
Elaine Pimentel

Henri Prade
David Pym
Ruy de Queiroz
Ram Ramanujam
Chrtian Retoré
Ulrike Sattler
Jörg Siekmann
Jane Spurr
Kaile Su
Leon van der Torre
Yde Venema
Rineke Verbrugge
Heinrich Wansing
Jef Wijsen
John Woods
Michael Wooldridge
Anna Zamansky

Scope and Submissions

This journal considers submission in all areas of pure and applied logic, including:

- pure logical systems
- proof theory
- constructive logic
- categorical logic
- modal and temporal logic
- model theory
- recursion theory
- type theory
- nominal theory
- nonclassical logics
- nonmonotonic logic
- numerical and uncertainty reasoning
- logic and AI
- foundations of logic programming
- belief revision
- systems of knowledge and belief
- logics and semantics of programming
- specification and verification
- agent theory
- databases
- dynamic logic
- quantum logic
- algebraic logic
- logic and cognition
- probabilistic logic
- logic and networks
- neuro-logical systems
- complexity
- argumentation theory
- logic and computation
- logic and language
- logic engineering
- knowledge-based systems
- automated reasoning
- knowledge representation
- logic in hardware and VLSI
- natural language
- concurrent computation
- planning

This journal will also consider papers on the application of logic in other subject areas: philosophy, cognitive science, physics etc. provided they have some formal content.

Submissions should be sent to Jane Spurr (jane.spurr@kcl.ac.uk) as a pdf file, preferably compiled in LaTeX using the IFCoLog class file.

Contents

ARTICLES

Editorial 2927
   Gabriella Pigozzi and Leendert van der Torre

Multiagent Deontic Logic and its Challenges from a Normative Systems
   Perspective 2929
   Gabriella Pigozzi and Leendert van der Torre

Detachment in Normative Systems:
   Examples, Inference Patterns, Properties 2995
   Xavier Parent and Leendert van der Torre

Handling Norms in Multi-agent Systems by Means of
   Formal Argumentation 3039
   *Célia da Costa Pereira, Beishui Liao, Alessandra Malerba, Antonino Rotolo, Andrea
   G. B. Tettamanzi, Leendert van der Torre and Serena Villata*

Logics for Games, Emotions and Institutions 3075
   Emiliano Lorini

EDITORIAL

GABRIELLA PIGOZZI
Université Paris-Dauphine, PSL Research University, CNRS, LAMSADE, 75016 Paris, France
gabriella.pigozzi@dauphine.fr

LEENDERT VAN DER TORRE
University of Luxembourg, Maison du Nombre, 6, Avenue de la Fonte, L-4364 Esch-sur-Alzette
leon.vandertorre@uni.lu

This special issue contains the journal version of four contributions to the Handbook of Normative Multi-Agent Systems (NorMAS), which will appear at College Publications. The NorMAS initiative aims at providing a comprehensive coverage of both the state of the art and future research perspectives in the interdisciplinary field of normative multi-agent systems. It is meant to be an open community effort and a service to current and future students and researchers interested in this field.

We invite the readers to buy the forthcoming handbook for a full view. Please visit the website for more information and feel free to send us comments, suggestions and proposals: http://normativemas.org/

The articles in this special issue and the chapters in the handbook give a survey of the area and may also contain a more personal view. For the survey part, at least the work reported in the NorMAS conference series is discussed. Instead of just a historical overview, the authors also address new developments, open topics and emerging areas. The handbooks appeal to all disciplines, including logic, computer science, law, philosophy, and linguistics. The articles in this special issue reflect the development of the logical analysis of normative multi-agent systems theory during the last two decades, with a special emphasis on the role played by deontic logic and normative systems. More information can be found in the handbook on deontic logic and normative systems, or the website http://deonticlogic.org/

In the first paper of the issue, Pigozzi and van der Torre give an overview of several challenges studied in deontic logic, with an emphasis on problems of multiagent deontic logic and problems related to normative systems. Fifteen challenges for multiagent deontic logic are considered, even though such list is by no means final. The three central concepts in these challenges are preference, agency, and norms.

In "Detachment in Normative Systems: Examples, Inference Patterns, Properties" Parent and van der Torre provide a systematic overview of detachment to compare traditional or standard semantics for deontic logic and alternative approaches. The focus is on inference patterns and proof-theory instead of semantical considerations. Despite the many frameworks for reasoning about rules and norms, and the many examples about detachment from normative systems, there are few properties to compare and analyse ways to detach obligations from rules and norms. The problem of detachment is addressed by surveying examples, inference patterns and properties from the deontic logic literature.

One approach used for the resolution of conflicting norms and norm compliance is formal argumentation. However, no comprehensive formal model of normative reasoning from arguments has been proposed yet. In "Handling Norms in Multi-Agent Systems by Means of Formal Argumentation" da Costa Pereira et al. present three challenges to illustrate the variety of applications of formal argumentation techniques in the field of normative multi-agent systems. Three examples from the literature of handling norms by means of formal argumentation are considered.

In the final paper of the issue, Lorini tackles the following question: how do institutions and norms, that are grounded on agents' collective attitudes retroactively influence decision-making and action? The discussion is organized around two main issues. The first one is the role of mental attitudes in decision-making and in action performance as well as the relationship between mental attitudes and emotions. The second issue considers how collective attitudes are generated from mental attitudes as well as the relationship between institutions and norms, on the one hand, and collective attitudes, on the other hand.

Multiagent Deontic Logic and its Challenges from a Normative Systems Perspective

Gabriella Pigozzi
Université Paris-Dauphine, PSL Research University, CNRS, LAMSADE, 75016 Paris, France
gabriella.pigozzi@dauphine.fr

Leendert van der Torre
University of Luxembourg, Maison du Nombre, 6, Avenue de la Fonte, L-4364 Esch-sur-Alzette
leon.vandertorre@uni.lu

Abstract

This article gives an overview of several challenges studied in deontic logic, with an emphasis on challenges involving agents. We start with traditional modal deontic logic using preferences to address the challenge of contrary-to-duty reasoning, and STIT theory addressing the challenges of non-deterministic actions, moral luck and procrastination. Then we turn to alternative norm-based deontic logics detaching obligations from norms to address the challenge of Jørgensen's dilemma, including the question how to derive obligations from a normative system when agents cannot assume that other agents comply with their norms. We discuss also some traditional challenges from the viewpoint of normative systems: when a set of norms may be termed 'coherent', how to deal with normative conflicts, how to combine normative systems and traditional deontic logic, how various kinds of permission can be accommodated, how meaning postulates and counts-as conditionals can be taken into account,

---

The authors thank Jan Broersen and Jörg Hansen for their joint work on earlier versions of some sections of this article, and Davide Grossi and Xavier Parent for insightful and helpful comments on a preliminary version of this article. The contribution of G. Pigozzi was supported by the Deutsche Forschungsgemeinschaft (DFG) and the Czech Science Foundation (GACR) as part of the joint project From Shared Evidence to Group Attitudes (RO 4548/6-1). This work is supported by the European Union's Horizon 2020 research and innovation programme under the Marie Curie grant agreement No: 690974 (Mining and Reasoning with Legal Texts, MIREL).

how sets of norms may be revised and merged, and how normative systems can be combined with game theory. The normative systems perspective means that norms, not ideality or preference, should take the central position in deontic semantics, and that a semantics that represents norms explicitly provides a helpful tool for analysing, clarifying and solving the problems of deontic logic. We focus on the challenges rather than trying to give full coverage of related work, for which we refer to the handbook of deontic logic and normative systems.[1]

Introduction

Deontic logic [116, 34] is the field of logic that is concerned with normative concepts such as obligation, permission, and prohibition. Alternatively, a deontic logic is a formal system capturing the essential logical features of these concepts. Typically, a deontic logic uses Op to mean that it is obligatory that p, (or it ought to be the case that p), and Pp to mean that it is permitted, or permissible, that p. The term 'deontic' is derived from the ancient Greek *déon*, meaning that "which is binding or proper".

Deontic logic can be used for reasoning about normative multiagent systems, i.e. about multiagent systems with normative systems in which agents can decide whether to follow the explicitly represented norms, and the normative systems specify how and to which extent agents can modify the norms [16, 6]. Normative multiagent systems need to combine normative reasoning with agent interaction, and thus raise the challenge to relate the logic of normative systems to game theory [109].

Traditional (or "standard") deontic logic is a normal propositional modal logic of type KD, which means that it extends the propositional tautologies with the axioms $K : O(p \to q) \to (Op \to Oq)$ and $D : \neg(Op \wedge O\neg p)$, and it is closed under the inference rules *modus ponens* $p, p \to q/q$ and *generalization* or *necessitation* p/Op. Prohibition and permission are defined by $Fp = O\neg p$ and $Pp = \neg O\neg p$. Traditional deontic logic is an unusually simple and elegant theory. An advantage of its modal-logical setting is that it can easily be extended with other modalities such as epistemic or temporal operators and modal accounts of action. In this article we illustrate the combination of deontic logic with a modal logic of action, called STIT logic [58].

Not surprisingly for such a highly simplified theory, there are many features of actual normative reasoning that traditional deontic logic does not capture. Noto-

---

[1] Sections 2-4 are based on a review of Horty's book on obligation and agency [23], Section 1 and Sections 5-14 are based on a technical report of a Dagstuhl seminar [52], and Section 15 is based on an article of the second author of this paper [109].

rious are the so-called 'paradoxes of deontic logic', which are usually dismissed as consequences of the simplifications of traditional deontic logic. For example, Ross's paradox [99] is the counterintuitive derivation of "you ought to mail or burn the letter" from "you ought to mail the letter." It is typically viewed as a side effect of the interpretation of 'or' in natural language.

In this article we discuss also an example of norm based semantics, called input/output logic, to discuss challenges related to norms and detachment. Maybe the most striking feature of the abstract character of traditional deontic logic is that it does not explicitly represent the norms of the system, only the obligations and permissions which can be detached from the norms in a given context. This is an obvious limitation when using deontic logic to reason about normative multiagent systems, in which norms are represented explicitly.

In this article we consider the following fifteen challenges for multiagent deontic logic. The list of challenges is by no means final. Other problems may be considered equally important, such as how a hierarchy of norms (or of the norm-giving authorities) is to be respected, how general abstract norms relate to individual concrete obligations, how norms can be interpreted, or how various kinds of imperatives can be distinguished. We do not consider deontic logics for specification and verification of multiagent systems [20, 1], but we focus on normative reasoning within multiagent systems. The three central concepts in these challenges are preference, agency, and norms. Regarding agency, we consider individual agent action as well as agent interaction in games.

1. Contrary-to-duty reasoning, preference and violation — preference
2. Non-deterministic actions: ought-to-do vs ought-to-be — agency
3. Moral luck and the driving example — agency
4. Procrastination: actualism vs possibilism — agency
5. Jørgensen's dilemma and the problem of detachment — norms
6. Multiagent detachment — norms
7. Coherence of a normative system — norms
8. Normative conflicts and dilemmas — preference & norms
9. Descriptive dyadic obligations and norms — preference & norms
10. Permissive norms — preference & norms
11. Meaning postulates and intermediate concepts — norms
12. Constitutive norms — norms
13. Revision of a normative system — norms
14. Merging normative systems — norms
15. Games, norms and obligations — norms & agency

To discuss these challenges, we repeat the basic definitions of so-called standard deontic logic, dyadic standard deontic logic, deontic STIT logic, and input/output logic. The article thus contains several definitions, but these are not put to work in any theorems or propositions, for which we refer to the handbook of deontic logic and normative systems [34]. The point of introducing formal definitions in this article is just to have a reference for the interested reader. Likewise, the interested reader should consult the handbook of deontic logic and normative systems for a more comprehensive description of the work done on each challenge, as in this article we can mention only a few references for each challenge.

1 Contrary-to-duty reasoning, preference and violation

In this section we discuss how the challenge of the contrary-to-duty paradoxes leads to traditional modal deontic logic introduced at the end of the sixties, based on dyadic operators and preference based semantics. Moreover, we contrast this use of preference in deontic logic with the use of preference in decision theory.

1.1 Chisholm's paradox

Suppose we are given a code of conditional norms, that we are presented with a condition (input) that is unalterably true, and asked what obligations (output) it gives rise to. It may happen that the condition is something that should not have been true in the first place. But that is now water under the bridge: we have to "make the best out of the sad circumstances" as B. Hansson [53] put it. We therefore abstract from the deontic status of the condition, and focus on the obligations that are consistent with its presence. How to determine this in general terms, and if possible in formal ones, is the well-known problem of contrary-to-duty conditions as exemplified by the notorious contrary-to-duty paradoxes. Chisholm's paradox [28] consists of the following four sentences:

(1) It ought to be that a certain man go to the assistance of his neighbours.
(2) It ought to be that if he does go, he tell them he is coming.
(3) If he does not go then he ought not to tell them he is coming.
(4) He does not go.

Furthermore, intuitively, the sentences derive the following sentence (5):

(5) He ought not to tell them he is coming.

Chisholm's paradox is a contrary-to-duty paradox, since it contains both a primary obligation to go, and a secondary obligation not to tell if the agent does not go. Traditionally, the paradox was approached by trying to formalise each of the sentences in an appropriate language of deontic logic. However, in traditional (or "standard") deontic logic, i.e. the normal propositional modal logic of type KD, it turned out that either the set of formulas is inconsistent, or one formula is a logical consequence of another formula. Yet intuitively the natural-language expressions that make up the paradox are consistent and independent from each other: this is why it is called a paradox. The problem is thus:

Challenge 1. *How do we reason with contrary-to-duty obligations which are in force only in case of norm violations?*

There are various kinds of scenarios which are similar to Chisholm's scenario. For example, there is a key difference between contrary-to-duties proper, and reparatory obligations, because the latter cannot be atemporal [98]. Though Chisholm presented his challenge as essentially a single agent decision problem, we can as well reformulate it as a multiagent reasoning problem:

(1) It is obligatory that i sees to it that p (i should do p).
(2) It is obligatory that j sees to it that q if i does not see to it that p
 (j should sanction i if i does not do as told).
(3) It is obligatory that j does not see to it that q if i sees to it that p
 (j should not sanction i if i does as told).
(4) i does not do as told.

The logic may give us the paradoxical conclusion that j should see to it that q and he should see to it that not q. For example, van Benthem, Grossi and Liu [108] give the following example, in the formulation proposed by Åqvist [7]:

(1) It ought to be that Smith refrains from robbing Jones.
(2) Smith robs Jones.
(3) If Smith robs Jones, he ought to be punished for robbery.
(4) It ought to be that if Smith refrains from robbing Jones he is not punished for robbery.

As explained in detail in the following subsections, the development of dyadic deontic operators as well as the introduction of temporally relative deontic logic operators can be seen as a direct result of Chisholm's paradox. Since the robbing takes place before the punishment, the example can quite easily be represented once time is made explicit [110]. If you make time explicit or you direct obligations to different agents, then the paradox disappears, in a way. However, both the fact that time and agency are present may distract from the key point behind the example. Therefore also atemporal, non-agency version of the paradox allow to address to the core challenge of the issue. For example, Prakken and Sergot [98] consider the following variant of Chisholm's scenario:

(1) It ought to be that there is no dog.
(2) If there is a dog, there should be a sign.
(3) If there is no dog, there should be no sign.
(4) There is a dog.

When a new deontic logic is proposed, the traditional contrary-to-duty examples are always the first benchmark examples to be checked. It may be observed here that some researchers in deontic logic doubt that contrary-to-duties can still be considered a challenge, because due to extensive research by now we know pretty much everything about them. The deontic logic literature is full of (at least purported) solutions. In other words, these researchers doubt that deontic logic still needs more research on contrary-to-duties. Indeed, it appears to be difficult to make an original contribution to this vast literature, but new twists are still identified [96].

1.2 Monadic deontic logic

Traditional or 'standard' deontic logic, often referred to as SDL, was introduced by Von Wright [116].

1.2.1 Language

Let Φ be a set of propositional letters. The language of traditional deontic logic $\mathfrak{L}_D$ is given by the following BNF:

$$\varphi := \bot \mid p \mid \neg\varphi \mid (\varphi \wedge \varphi) \mid \bigcirc\varphi \mid \Box\varphi$$

where $p \in \Phi$. The intended reading of $\bigcirc\varphi$ is "φ is obligatory" and the intended reading of $\Box\varphi$ is "φ is necessary". Moreover we use $P\varphi$, read as "φ is permitted", as an abbreviation of $\neg\bigcirc\neg\varphi$ and $F\varphi$, "φ is forbidden", as an abbreviation of $\bigcirc\neg\varphi$. Likewise, $\vee$, $\rightarrow$ and $\leftrightarrow$ are defined in the usual way.

1.2.2 Semantics

The semantics is based on an accessibility relation that gives all the ideal alternatives of a world.

Definition 1.1. *A deontic relational model $M = (W, R, V)$ is a structure where:*

- *W is a nonempty set of worlds.*
- *R is a serial relation over W. That is, $R \subseteq W \times W$ and for all $w \in W$, there exist $v \in W$ such that Rwv.*

- V is a valuation function that assigns a subset of W to each propositional letter p. Intuitively, $V(p)$ is the set of worlds in which p is true.

A formula $\bigcirc\varphi$ is true at world w when φ is true in all the ideal alternatives of w.

Definition 1.2. *Given a relational model M, and a world s in M, we define the satisfaction relation $M, s \models A$ ("world s satisfies A in M") by induction on A using the clauses:*

- $M, s \models p$ iff $s \in V(p)$.
- $M, s \models \neg\varphi$ iff not $M, s \models \varphi$.
- $M, s \models (\varphi \wedge \psi)$ iff $M, s \models \varphi$ and $M, s \models \psi$.
- $M, s \models \bigcirc\varphi$ iff for all t, if Rst then $M, t \models \varphi$.
- $M, s \models \Box\varphi$ iff for all $t \in W$, $M, t \models \varphi$.

For a set Γ of formulas, we write $M, s \models \Gamma$ iff for all $\varphi \in \Gamma$, $M, s \models \varphi$. For a set Γ of formulas and a formula φ, we say that φ is a consequence of Γ (written as $\Gamma \models \varphi$) if for all models M and all worlds $s \in W$, if $M, s \models \Gamma$ then $M, s \models \varphi$.

1.2.3 Limitations

The following example is a variant of the scenario originally phrased by Chisholm in 1963. There is widespread agreement in the literature that, from the intuitive point of view, this set of sentences is consistent, and its members are logically independent of each other.

(A) It ought to be that Jones does not eat fast food for dinner.

(B) It ought to be that if Jones does not eat fast food for dinner, then he does not go to McDonald's.

(C) If Jones eats fast food for dinner, then he ought to go to McDonald's.

(D) Jones eats fast food for dinner.

Below are three ways to formalise this example. The first attempt is inconsistent. The second attempt is redundant due to $\bigcirc\neg f \models \bigcirc(f \to m)$. The third attempt is redundant due to $f \models \neg f \to \bigcirc\neg m$.

(A_a)	$\bigcirc\neg f$	(A_b)	$\bigcirc\neg f$	(A_c)	$\bigcirc\neg f$
(B_a)	$\bigcirc(\neg f \to \neg m)$	(B_b)	$\bigcirc(\neg f \to \neg m)$	(B_c)	$\neg f \to \bigcirc\neg m$
(C_a)	$f \to \bigcirc m$	(C_b)	$\bigcirc(f \to m)$	(C_c)	$f \to \bigcirc m$
(D_a)	f	(D_b)	f	(D_c)	f

However, it is not very hard to meet the two requirements of consistency and logical independence. The following representation is an example. It comes with apparently strong assumptions, because B_1/C_1 seem to say that my (conditional) obligations are necessary. For instance, Anderson argued that norms are contingent, because we make our rules; they are not (logical) necessities. However, we could also say that the $\Box$ is just part of the definition of a strict conditional. Also, we could represent the first obligation as $\Box \bigcirc \neg f$.

(A_1) $\quad \bigcirc \neg f$
(B_1) $\quad \Box(\neg f \rightarrow \bigcirc \neg m)$
(C_1) $\quad \Box(f \rightarrow \bigcirc m)$
(D_1) $\quad \neg f$

More seriously, a drawback of the SDL representation $A_1 - D_1$ is that it does not represent that ideally, the man does not eat fast food and does not go to McDonald's. In the ideal world, Jones goes to McDonald, yet he does not eat fast food. Moreover, there does not seem to be a similar solution for the following variant of the scenario. It is a variant of Forrester's paradox [33], also known as the gentle murderer paradox: You should not kill, but if you kill, you should do it gently.

(AB) It ought to be that Jones does not eat fast food and does not go to McDonald's.

(C) If Jones eats fast food, then he ought to go to McDonald's.

(D) Jones eats fast food for dinner.

Moreover, SDL uses a binary classification of worlds into ideal/non-ideal, whereas many situations require a trade-off between violations. The challenge is to extend the semantics of SDL in order to overcome this limitation. For example, one can add distinct modal operators for primary and secondary obligations, where a secondary obligation is a kind of reparational obligation. From $A_2 - D_2$ we can derive only $\bigcirc_1 m \wedge \bigcirc_2 \neg m$, which is perfectly consistent.

(A_2) $\quad \bigcirc_1 \neg f$
(B_2) $\quad \bigcirc_1(\neg f \rightarrow \neg m)$
(C_2) $\quad f \rightarrow \bigcirc_2 m$
(D_2) $\quad f$

However, it may not always be easy to distinguish primary from secondary obligations, because it may depend on the context whether an obligation is primary or secondary. For example, if we leave out **A**, then **C** would be a primary obligation instead of a secondary one. Carmo and Jones [25] therefore put as an additional requirement for a solution of the paradox that **B** and **C** are represented in the same way (as in A_1-D_1). Also, the distinction between $\bigcirc_1$ and $\bigcirc_2$ is insufficient for extensions of the paradox that seem to need also operators like $\bigcirc_3$, $\bigcirc_4$, etc, such as the following **E** and **F**.

(E) If Jones eats fast food but does not go to McDonald's, then he should go to Quick.

(F) If Jones eats fast food but does not go to McDonald's or to Quick, then he should ...

1.2.4 SDL proof system

The proof system of traditional deontic logic Λ_D is the smallest set of formulas of $\mathcal{L}_D$ that contains all propositional tautologies, together with the following axioms:

K $\bigcirc(\varphi \to \psi) \to (\bigcirc\varphi \to \bigcirc\psi)$

D $\bigcirc\varphi \to P\varphi$

and is closed under *modus pones*, and *generalization* (that is, if $\varphi \in \Lambda_D$, then $\bigcirc\varphi \in \Lambda_D$).

For every $\varphi \in \mathcal{L}_D$, if $\varphi \in \Lambda_D$ then we say φ is a theorem and write $\vdash \varphi$. For a set of formulas Γ and formula φ, we say φ is deducible form Γ (write $\Gamma \vdash \varphi$) if $\vdash \varphi$ or there are formulas $\psi_1, \ldots, \psi_n \in \Gamma$ such that $\vdash (\psi_1 \wedge \ldots \wedge \psi_n) \to \varphi$.

1.3 Dyadic deontic logic

Inspired by rational choice theory in the sixties, preference-based semantics for traditional deontic logic was used by, for example, Danielsson [32], Hansson [53], van Fraassen [115], Lewis [74], and Spohn [104]. The obligations of Chisholm's paradox can be represented by a preference ordering, like:

$$\neg f \wedge \neg m > \neg f \wedge m > f \wedge m > f \wedge \neg m$$

Extensions like **E** and **F** can be incorporated by further refining the preference relation. The language is extended with dyadic operators $\bigcirc(p|q)$, which is true iff the preferred q worlds satisfy p. The class of logics is called Dyadic 'Standard' Deontic Logic or DSDL. The notation is inspired by the representation of conditional probability.

1.3.1 Language

Given a set Φ of propositional letters. The language of DSDL $\mathfrak{L}_D$ is given by the following BNF:
$$\varphi := \bot \mid p \mid \neg\varphi \mid (\varphi \wedge \varphi) \mid \Box\varphi \mid \bigcirc(\varphi|\varphi)$$

The intended reading of $\Box\varphi$ is "necessarily φ", $\bigcirc(\varphi|\psi)$ is "It ought to be φ, given ψ". Moreover we use $P(\varphi|\psi)$, read as "φ is permitted, given ψ", as an abbreviation of $\neg\bigcirc(\neg\varphi|\psi)$, and $\Diamond\varphi$, read as "possibly φ", as an abbreviation of $\neg\Box\neg\varphi$.

Unconditional obligations are defined in terms of the conditional ones by $\bigcirc p = \bigcirc(p|\top)$, where $\top$ stands for any tautology.

1.3.2 Semantics

The semantics is based on an accessibility relation that gives all better alternatives of a world.

Definition 1.3. *A preference model $M = (W, \geq, V)$ is a structure where:*

- *W is a nonempty set of worlds.*

- *$\geq$ is a reflexive, transitive relation over W satisfying the following limitedness requirement: if $||\varphi|| \neq \emptyset$ then $\{x \in ||\varphi|| : (\forall y \in ||\varphi||) x \geq y\} \neq \emptyset$. Here $||\varphi|| = \{x \in W : M, x \vDash \varphi\}$.*

- *V is a standard propositional valuation such that for every propositional letter p, $V(p) \subseteq W$.*

Definition 1.4. *Formulas of $\mathfrak{L}_D$ are interpreted in preference models.*

- *$M, s \vDash p$ iff $s \in V(p)$.*
- *$M, s \vDash \neg\varphi$ iff not $M, s \vDash \varphi$.*
- *$M, s \vDash (\varphi \wedge \psi)$ iff $M, s \vDash \varphi$ and $M, s \vDash \psi$.*
- *$M, s \vDash \Box\varphi$ iff $\forall t \in W, M, t \vDash \varphi$.*
- *$M, s \vDash \bigcirc(\psi|\varphi)$ iff $\forall t(((M, t \vDash \varphi) \& \forall u(M, u \vDash \varphi) \Rightarrow t \geq u) \Rightarrow M, t \vDash \psi)$.*

Intuitively, $\bigcirc(\psi|\varphi)$ holds whenever the best φ-worlds are ψ-worlds. The Chisholm's scenario can be formalised in DSDL as follows:

(A_3) $\bigcirc \neg f$

(B_3) $\bigcirc(\neg m|\neg f)$

$(C_3)\ \bigcirc(m|f)$

$(D_3)\ f$

A challenge of both the multiple obligation solution using $\bigcirc_1, \bigcirc_2, \ldots$ and the preference based semantics is to combine preference orderings, for example combining the Chisholm preferences with preferences originating from the Good Samaritan paradox:

(AB') A man should not be robbed.

(C') If he is robbed, he should be helped.

(D') A man is robbed.

$$\neg r \wedge \neg h > r \wedge h > r \wedge \neg h$$

The main drawback of DSDL is that in a monotonic setting, we cannot detach the obligation $\bigcirc m$ from the four sentences. In fact, the preference based solution represents **A**, **B** and **C**, but has little to say about **D**. So the dyadic representation $A_3 - D_3$ highlights the dilemma between factual detachment (FD) and deontic detachment (DD). We cannot have both FD and DD, as we derive a dilemma $\bigcirc \neg m \wedge \bigcirc m$.

$$\frac{\bigcirc(m|f), f}{\bigcirc m} FD \qquad \frac{\bigcirc(\neg m|\neg f), \bigcirc \neg f}{\bigcirc \neg m} DD$$

1.3.3 DSDL proof system

The proof system of traditional deontic logic Λ_D, also referred as Aqvist's system G, is the smallest set of formulas of $\mathcal{L}_D$ that contains all propositional tautologies, the following axioms. The names of the labels are taken from Parent [93]:

S5 S5-schemata for $\Box$

COK $\bigcirc(B \to C|A) \to (\bigcirc(B|A) \to \bigcirc(C|A))$

Abs $\bigcirc(B|A) \to \Box\bigcirc(B|A)$

CON $\Box B \to \bigcirc(B|A)$

Ext $\Box(A \leftrightarrow B) \to (\bigcirc(C|A) \leftrightarrow \bigcirc(C|B))$

Id $\bigcirc(A|A)$

C $\bigcirc(C|(A \wedge B)) \to \bigcirc((B \to C)|A)$

D $\star$ $\Diamond A \to (\bigcirc(B|A) \to P(B|A))$

S $(P(B|A) \wedge \bigcirc((B \to C)|A)) \to \bigcirc(C|(A \wedge B))$

and is closed under *modus ponens*, and *generalization* (that is, if $\varphi \in \Lambda_D$, then $\Box\varphi \in \Lambda_D$).

1.3.4 The use of preferences in decision theory

Arrow's condition of rational choice theory says that if C are the best alternatives of A, and $B \cap C$ is nonempty, then $B \cap C$ are the best alternatives of $A \cap B$. This principle is reflected by the S axiom of DSDL:

$$(P(B|A) \wedge \bigcirc((B \to C)|A)) \to \bigcirc(C|(A \wedge B))$$

Moreover, we may represent a preference or comparative operator $\succ$ in the language, and define the dyadic operator in terms of the preference logic:

$$O(\psi \mid \phi) =_{def} (\phi \wedge \psi) \succ (\phi \wedge \neg\psi)$$

One may wonder whether the parallel between deontic reasoning and rational choice can be extended to utility theory, decision theory, game theory, planning, and so on. First, consider a typical example from Prakken and Sergot's Cottage Regulations [98]: there should be no fence, if there is a fence there should be a white fence, if there is a non-white fence, it should be black, if there is a fence which is neither white nor black, then .... This part of the cottage regulations is related to Forrester's paradox [33]. However, note the following difference between Forrester's paradox and the cottage regulations. Once you kill someone, it can no longer be undone, whereas if you build a fence, you can still remove it. The associated preferences of the fence example are:

no fence > white fence > black fence > ...

If this represents a utility ordering over states, then we miss the representation of action [97]. For example, it may be preferred that the sun shines, but we do not say that the sun should shine. As a simple model of action, one might distinguish controllable from uncontrollable propositions [19], and restrict obligations to controllable propositions. Moreover, we may consider actions instead of states: we should remove the fence if there is one, we may paint the fence white, we may paint it black, etc.

remove > paint white > paint black > ...

We may interpret this preference ordering as an ordering of expected utility of actions. Alternatively, the ordering may be generated by another decision rule, such as maximin or minimal regret. Once we are working with a decision theoretic semantics, we may represent probabilities explicitly, or model causality. For example, let n stand for not doing homework and g for getting a good grade for a test. Then we may have the following preference order, which does not reflect that doing homework causes good grades:

$$n \wedge g > \neg n \wedge g > n \wedge \neg g > \neg n \wedge \neg g$$

1.3.5 The use of goals in planning and agent theory

We may interpret $O\phi$ or $O(\phi \mid \psi)$ as goals for ϕ, rather than obligations. This naturally leads to the distinction between maintenance and achievement goals, and to extensions of the logic with beliefs and intentions. Belief-Desire-Intention or BDI logics have been developed as formalizations of BDI theory.

BDI theory is developed in the theory of mind and has been based on folk psychology. In planning, more efficient alternatives to classical planning have been developed, for example based on hierarchical or graph planning.

The following example is a more challenging variant of Chisholm's scenario using anankastic conditionals [31], also known as hypothetical imperatives. The four sentences can be given a consistent interpretation, when the second sentence is interpreted as a classical conditional, and the third sentence is interpreted as an anankastic conditional.

(a) It ought to be that you do not smoke.

(b) If you want to smoke, then you should not buy cigarettes.

(c) If you want to smoke, then you should buy cigarettes.

(d) You want to smoke.

1.4 Defeasible Deontic Logic: detachment and constraints

Defeasible deontic logics (DDLs) use techniques developed in non-monotonic logic, such as constrained inference [60, 86]. Using these techniques, we can derive Om from only the first two sentences **A** and **B**, but not from all four sentences **A-D**. Consequently, the inference relation is not monotonic. For example, we may read $O(\phi|\psi)$ as follows: if the facts are exactly ψ, then ϕ is obligatory. This implies that we no longer have that $O(\phi)$ is represented by $O(\phi|\top)$.

In a similar fashion, in deontic update semantics (see van der Torre and Tan [111, 113, 112]) facts are updates that restrict the domain of the model. They make a fact 'settled' in the sense that it will never change again even after future updates of the same sort. Van Benthem et al. [108] use dynamic logic to phrase such a dynamic approach within standard modal logic including reduction axioms and standard model theory. They rehabilitate classical modal logic as a legitimate tool to do deontic logic, and position deontic logic within the growing dynamic logic literature.

A drawback of the use of non-monotonic techniques is that we often have that violated obligations are no longer derived. This is sometimes referred to as the drowning problem. For example, in the cottage regulations, if it is no longer derived that there should be no fence once there is a fence, then how do we represent that a violation has occurred?

A second related drawback of this solution is that it does not give the cue for action that the decision maker should change his mind. For example, once there is a fence, it does not represent the obligation to remove the fence.

A third drawback of this approach is that the use of non-monotonic logic techniques like constraints should also be used to represent exceptions, and it thus raises the challenge how to distinguish violations from exceptions. This is highlighted by Prakken and Sergot's cottage regulations [98].

(A") It ought to be that there is no fence around the cottage.

(BC") If there is a fence around the cottage, then it ought to be white.

(G") If the cottage is close to a cliff, then there ought to be a fence.

(D") There is a fence around the cottage, which is close to a cliff.

We say more about defeasible deontic logic in Section 8.

1.5 Alternative approaches

Carmo and Jones [25] suggest that the representation of the facts is challenging, instead of the representation of the norms. In their approach, depending on the formalisation of the facts various obligations can be detached.

Another approach to Chisholm's paradox is to detach both obligations of the dilemma $\bigcirc \neg m \wedge \bigcirc m$, and represent them consistently using some kind of minimal deontic logic, for example using techniques from paraconsistent logic. From a practical reasoning point of view, a drawback of this approach is that a dilemma is not very useful as a moral cue for action. Moreover, intuitively it is not clear that the example presents a true dilemma. We say more about dilemmas in Section 9.

A recent representation of Chisholm's paradox [94, 95, 107] is to replace deontic detachment by so-called aggregative deontic detachment (ADD), and to derive from

A-D the obligation $\bigcirc(\neg f \wedge \neg m)$ and $\bigcirc m$, but not $\bigcirc \neg m$.

$$\frac{\bigcirc(m|f), f}{\bigcirc m} FD \qquad \frac{\bigcirc(\neg m|\neg f), \bigcirc \neg f}{\bigcirc(\neg m \wedge \neg f)} ADD$$

A possible drawback of these approaches is that we can no longer accept the principle of weakening (also known as inheritance).

$$\frac{\bigcirc(\neg m \wedge \neg f | \top)}{\bigcirc(\neg m | \top)} W$$

2 Non-deterministic actions: ought-to-do vs ought-to-be

We now turn to three specific challenges on agency and obligation, discussed in much more detail by Horty [58, 23]. His textbook is a prime reference for the use of deontic logic for multiagent systems. The central challenge Horty addresses is whether ought-to-do can be reduced to ought-to-be. A particular problem is the granularity of actions in case of non-deterministic effects, like flipping a coin or throwing a dice.

Challenge 2. *How to define obligations to perform non-deterministic actions?*

At first sight, we may define an obligation to do an action as an obligation that such an action is done, and we can thus reuse SDL or DSDL to define obligations regarding non-deterministic actions. In other words, it may seem that we can reduce ought-to-do to ought-to-be. However, as we discuss in Section 2.2, such a reduction is problematic. To explain this challenge, we first introduce a logic to express non-deterministic actions, so-called See-To-It-That or STIT logic.

2.1 Horty's STIT logic

We give a very brief overview of the main concepts of Horty's STIT logic. For more details and motivation we refer to Horty's textbook on obligation and agency [58]. As illustrated in Figure 1, a STIT model is a tree where each moment is a partitioning of traces or histories, where the partitioning $Choice_\alpha^m$ represents the choices of the agent at that moment. Each alternative of the choice is called an action K_1^m, K_2^m, etc. With each history a utility value is associated, and the higher the utility value, the better the history.

Formulas are evaluated with respect to moment-history pairs. Some typical formulas of Horty's utilitarian STIT-formalism are A, FA, $[\alpha \ cstit : A]$, and $\bigcirc A$

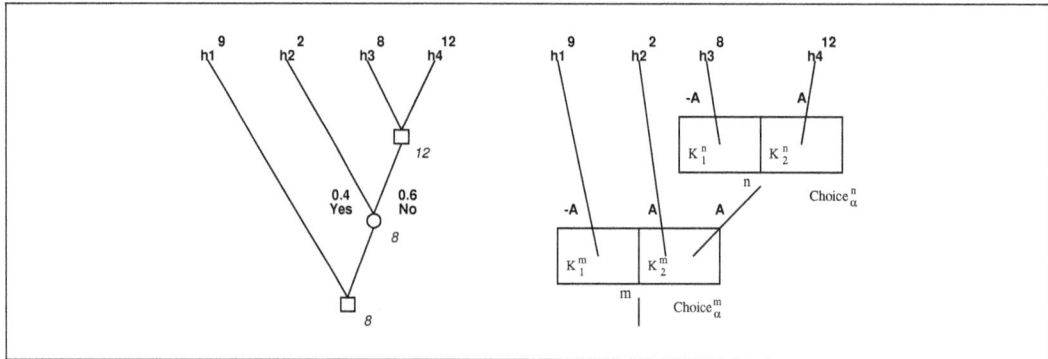

Figure 1: A decision tree and the corresponding utilitarian STIT-model

for 'the atomic proposition A', 'some time in the future A will be the case', 'agent α Sees To It That A', and 'it ought to be that A', respectively.

A is true at a moment-history pair m, h if and only if it is assigned the value true in the STIT-model, FA is true at a moment-history pair m, h if and only if there is some future moment on the history where A is true, $[\alpha\ cstit : A]$ is true at a moment history pair m, h if and only if A is true at all moment-history pairs through m that belong to the same *action* as m, h, and $\bigcirc A$ is true at a moment history pair m, h if and only if there is some history h' through m such that A is true at all pairs m, h'' for which the history h'' has a utility at least as high as h' ('moment determinate').

This semantic condition for the STIT-ought is a utilitarian generalisation of the standard deontic logic view (SDL) that 'it ought to be that A' means that A holds in all deontically optimal worlds.

On the STIT-model of Figure 1 we have $\mathcal{M}, m, h_3 \models A$ (directly from the valuation of atomic propositions on moment-history pairs), $\mathcal{M}, m, h_3 \models F\neg A$ (the proposition $\neg A$ is true later on, at moment n, on the history h_3 through m).

Also we have $\mathcal{M}, m, h_3 \models [\alpha\ cstit : A]$, because A holds for all histories through moment m belonging to the same action as h_3 (i.e. action K_2^m). Regarding ought-formulas we have: $\mathcal{M}, m, h_3 \models \bigcirc A$ and $\mathcal{M}, m, h_3 \models \bigcirc[\alpha\ cstit : A]$.

These two propositions are true for the same reason: the history h_4 through m has the highest utility (which means that we do not have to check conditions for histories with even higher utility) and satisfies both A and $[\alpha\ cstit : A]$ at m.

2.2 Gambling problem

Horty argues that ought-to-do statements are not just special kinds of ought-to-be statements. In particular, he claims that 'agent α ought to see to it that A' cannot be modelled by the formula $\bigcirc[\alpha\ cstit : A]$ ('it ought to be that agent α sees to it that A').

Justification of this claim is found in the 'gambling example'. This example concerns the situation where an agent faces the choice between gambling to double or lose five dollar (action K_1) and refraining from gambling (action K_2). This situation is sketched in the figure below.

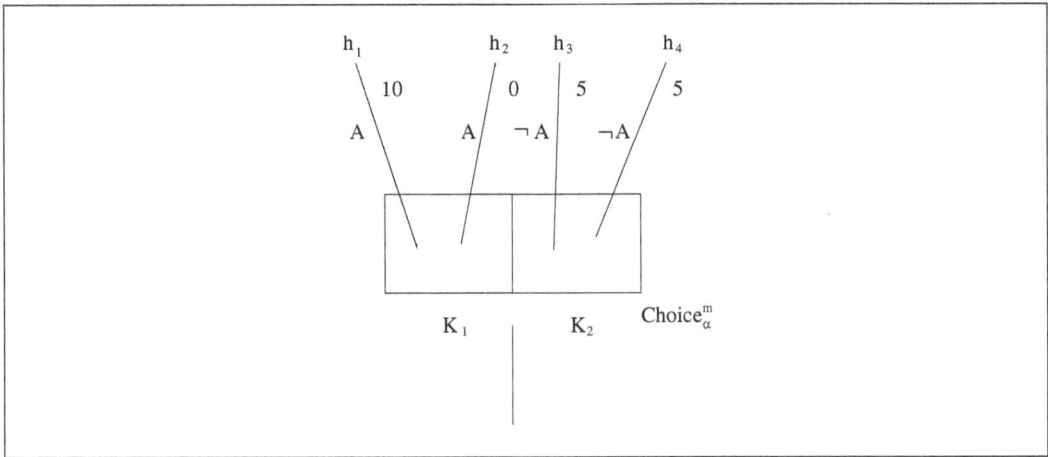

Figure 2: The gambling problem

The two histories that are possible by choosing action K_1 represent ending up with ten dollar by gaining five, and ending up with nothing by loosing all, respectively.

Also for action K_2, the game event causes histories to branch. But, for this action the two branches have equal utilities because the agent is not taking part in the game, thereby preserving his 5 dollar. Note this points to redundancy in the model representation: the two branches are logically indistinguishable, because there is no formula whose truth value would change by dropping one of them.

$\bigcirc[\alpha\ cstit : A]$ is true at m for history h_1 and for all histories with a higher utility (i.e. none), the formula $[\alpha\ cstit : A]$ is true. However, a reading of $\bigcirc[\alpha\ cstit : A]$ as 'agent α ought to perform action K_1' is counter-intuitive for this example. From the description of the gambling scenario it does not follow that one action is better than the other. In particular, without knowing the odds (the probabilities), we cannot say anything in favor of action K_1: by choosing it, we may either end up with more or with less utility than by doing K_2. The only thing one may observe is that action K_1 will be preferred by more adventurous agents. But that is not something the logic is concerned with.

This demonstrates that 'agent α ought to see to it that A' cannot be modelled by $\bigcirc[\alpha\ cstit : A]$. The cause of the mismatch can be explained as follows. Adapting and generalising the main idea behind SDL to the STIT-context, ought-to-be statements concern truth in a set of optimal histories ('worlds' in SDL). Optimality is directly determined by the utilities associated with individual histories. If ought-to-be is about optimal histories, then ought-to-do is about optimal actions. But, since actions are assumed to be non-deterministic, actions do not correspond with individual histories, but with *sets* of histories. This means that to apply the idea of optimality to the definition of ought-to-do operators, we have to generalise the notion of optimality such that it applies to *sets* of histories, namely, the sets that make up non-deterministic actions. More specifically, we have to *lift* the ordering of histories to an ordering of actions. The ordering of actions suggested by Horty is very simple: an action is strictly better than another action if all of its histories are at least as good as any history of the other action, and not the other way around.

Having lifted the ranking of histories to a ranking of actions, the utilitarian ought conditions can now be applied to actions. Thus, Horty defines the new operator 'agent α ought to see to it that A (in formula form: $\odot[\alpha\ cstit : A]$)' as the condition that for all actions not resulting in A there is a higher ranked action that does result in A, plus that all actions that are ranked even higher also result in A. This 'solves' the gambling problem. We do not have $\odot[\alpha\ cstit : A]$ or $\odot[\alpha\ cstit : \neg A]$ in the gambling scenario, because in the ordering of actions, K_1 is not better or worse than K_2.

3 Moral luck and the driving example

The gambling problem may be seen as a kind of moral luck: whether we obtain the utility of 10 or 0 is not due to our actions, but due to luck. The issue of moral luck is even more interesting in the case of multiple agents, where it depends on the actions of other agents whether you get utility 10 or 0.

Challenge 3. *How to deal with moral luck in normative reasoning?*

The driving example [58, p.119-121] is used to illustrate the difference between so-called dominance act utilitarianism and orthodox perspective on the agent's ought. Roughly, dominance act utilitarianism is that α ought to see to it that A just in case the truth of A is guaranteed by each of the optimal actions available to the agent—formally, that $\odot[\alpha\ cstit : A]$ should be settled true at a moment m just in case $K \subseteq |A|_m$ for each $K \in Optimal_\alpha^m$. When we adopt the orthodox perspective, the truth or falsity of ought statements can vary from index to index. The orthodox perspective is that α should see to it that A at a certain index just in case the truth of A is guaranteed by each of the actions available to the agent that are optimal given the circumstances in which he finds himself at this index.

"In this example, two drivers are travelling toward each other on a one-lane road, with no time to stop or communicate, and with a single moment at which each must choose, independently, either to swerve or to continue along the road. There is only one direction in which the drivers might swerve, and so a collision can be avoided only if one of the drivers swerves and the other does not; if neither swerves, or both do, a collision occurs. This example is depicted in Figure 3, where α and β represent the two drivers, K_1 and K_2 represent the actions available to α of swerving or staying on the road, K_3 and K_4 likewise represent the swerving or continuing actions available to β, and m represents the moment at which α and β must make their choice. The histories h_1 and h_3 are the ideal outcomes, resulting when one driver swerves and the other one does not; collision is avoided. The histories h_2 and h_4, resulting either when both drivers swerve or both continue along the road, represent non-ideal outcomes; collision occurs. The statement A, true at h_1 and h_2, expresses the proposition that α swerves." [58, p.119]

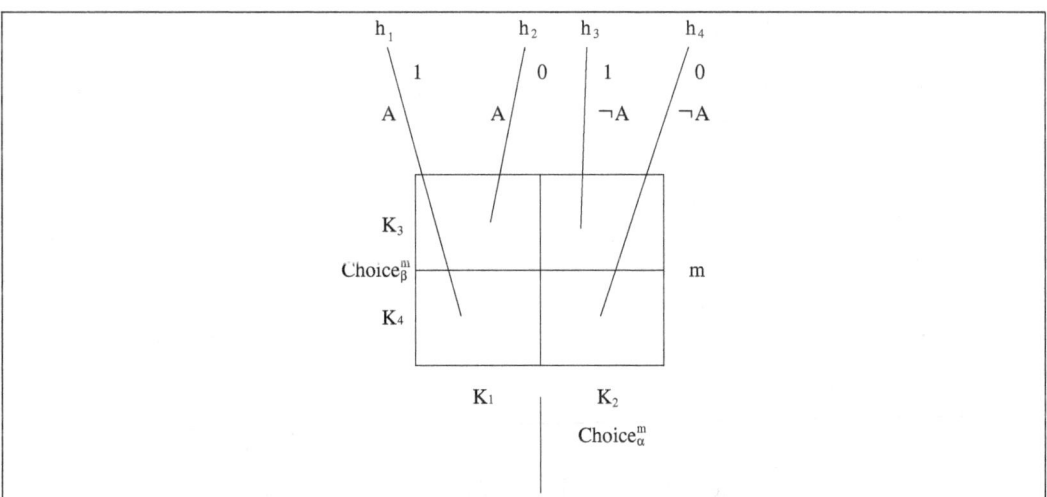

Figure 3: The driving example and moral luck

From the dominance point of view both actions available to α are classified as optimal, written as $Optimal_\alpha^m = \{K_1, K_2\}$. One of the optimal actions available to

α guarantees the truth of A and the other guarantees the truth of $\neg A$. Consequently $M, m \not\models \odot[\alpha \; cstit : A]$ and $M, m \not\models \odot[\alpha \; cstit : \neg A]$. From the orthodox point of view, we have $M, m, h_1 \models \bigcirc[\alpha \; cstit : A]$ and $M, m, h_2 \models \bigcirc[\alpha \; cstit : \neg A]$. What α ought to do at an index depends on what β does.

Horty concludes that from the standpoint of intuitive adequacy, the contrast between the orthodox and dominance deontic operators provides us with another perspective on the issue of moral luck, the role of external factors in our moral evaluations [58, p.121]. The orthodox ought is the one who after the actual event looks back to it. For example, when there has been a collision then α might say—perhaps while recovering from the hospital bed—that he ought to have swerved. The dominance ought is looking forward. Though the agent may legitimately regret his choice, it is not one for which he can be blamed, since either choice, at the time, could have led to a collision.

4 Procrastination: actualism vs possibilism

Practical reasoning is intimately related to reasoning about time. For example, if you are obliged and willing to visit a relative, but you always procrastinate this visit, then we may conclude that you violated this obligation. In other words, each obligation to do an action should come with a deadline [22, 11].

Challenge 4. *How to deal with procrastination in normative reasoning?*

The example of Procrastinate's choices [58, p. 162] illustrates the notion of strategic oughts. A strategy is a generalized action involving a series of actions. Like an action, a strategy determines a subset of histories. The set of admissible histories for a strategy σ is denoted $Adh(\sigma)$.

A crucial new concept here is the concept of a *Field*, which is basically a subtree of the STIT model which denotes that the agent's reasoning is limited to this range. A strategic ought is defined analogous to dominance act utilitarianism, in which action is replaced by strategy in a field. α ought to see to it that A just in case the truth of A is guaranteed by each of the optimal strategies available to the agent in the field—formally, that $\odot[\alpha \; cstit : A]$ should be settled true at a moment m just in case $Adh(\sigma) \subseteq |A|_m$ for each $\sigma \in Optimal_\alpha^m$. Horty observes some complications, and that a 'proper treatment of these issues might well push us beyond the borders of the current representational formalism' [p.150].

Horty also uses the example of Procrastinate's choices to distinguish between actualism and possibilism, for which he uses the strategic oughts, and in particular the notion of a field. Roughly, actualism is the view that an agent's current actions are to be evaluated against the background of the actions he is actually going to perform in the future. Possibilism is the view that an agent's current actions are to be evaluated against the background of the actions that he might perform in the future, the available future actions.

The example is due to Jackson and Pargetter [63].

> "Professor Procrastinate receives an invitation to review a book. He is the best person to do the review, has the time, and so on. The best thing that can happen is that he says yes, and then writes the review when the book arrives. However, suppose it is further the case that were to say yes, he would not in fact get around to writing the review. Not because of incapacity or outside interference or anything like that, but because he would keep on putting the task off. (This has been known to happen.) This although the best thing that can happen is for Procrastinate to say yes and then write, and he *can* do exactly this, what *would* happen in fact were he to say yes is that he would not write the review. Moreover, we may suppose, this latter is the worst thing which may happen.
>
> [...]
>
> According to possibilism, the fact that Procrastinate would not write the review were he to say yes is irrelevant. What matters is simply what is possible for Procrastinate. He can say yes and then write; that is best; that requires *inter alia* that he says yes; therefore, he ought to say yes. According to actualism, the fact that Procrastinate would not actually write the review were he to say yes is crucial. It means that to say yes would be in fact to realize the worst. Therefore, Procrastinate ought to say no."

Horty represents the example by the STIT model in Figure 4. Here, m_1 is the moment at which Procrastinate, represented as the agent α, chooses whether or not to accept the invitation: K_1 represents the choice of accepting, K_2 the choice of declining. If Procrastinate accepts the invitation, he then faces at m_2 the later choice of writing the review or not: K_3 represents the choice of writing the review, K_4 another choice that results in the review not being written. For convenience, Horty also supposes that at m_3 Procrastinate has a similar choice whether or not to write the review: K_5 represents the choice of writing, K_6 the choice of not writing. The history h_1, in which Procrastinate accepts the invitation and then writes the review, carries the greatest value of 10; the history h_2, in which Procrastinate accepts the invitation and then neglects the task, the least value of 0; the history h_4, in which he declines, such that a less competent authority reviews the book, carries an intermediate value of 5; and the peculiar h_3, in which he declines the invitation but then reviews the book anyway, carries a slightly lower value of 4, since he wastes his time, apart from doing no one else any good. The statement A represents the proposition that he accepts the invitation; the statement B represents the proposition that Procrastinate will write the review.

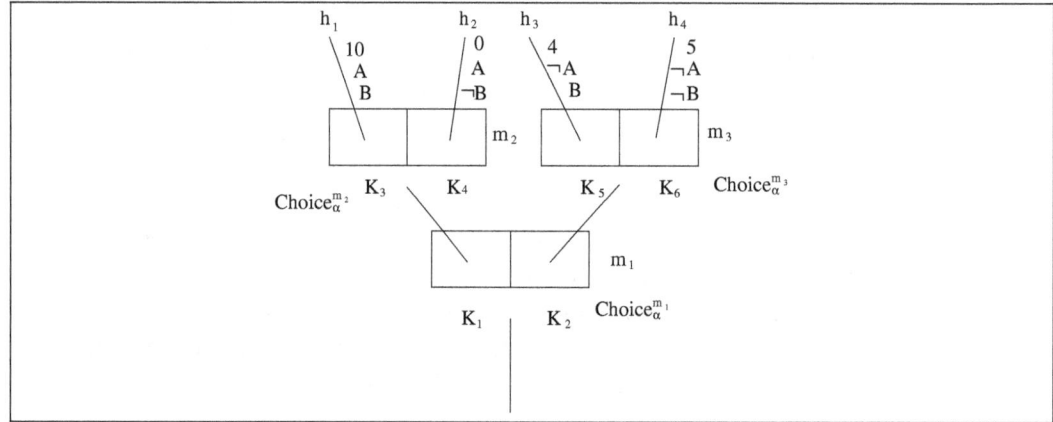

Figure 4: Procrastinate's choices

Now, in the possibilist interpretation, $M = \{m_1, m_2, m_3\}$ is the background field. In this interpretation, Procrastinate ought to accept the invitation because this is the action determined by the best available strategy—first accepting the invitation, and then writing the review. Formally, we have $Optimal_\alpha^M = \{\sigma_6\}$ with $\sigma_6 = \{\langle m_1, K_1\rangle, \langle m_2, K_3\rangle\}$. Since $Adh(\sigma_6) \subseteq |A|_m$, the strategic ought statement $\odot[\alpha\ cstit : A]$ is settled true in the field M. In the actualist interpretation, the background field may be narrowed to the set $M' = \{m_1\}$, which shifts from the strategic to the momentary theory of oughts. In this case, we have $\odot[\alpha\ cstit : A]$ is settled false. It is as if we choose to view Procrastinate as gambling on his own later choice in deciding whether to accept the invitation. However, from this perspective, this should not be viewed as a gamble; an important background assumption—and the reason that he should decline the invitation—is that he will not, in fact, write the review.

5 Jørgensen's dilemma and the problem of detachment

A philosophical problem that has had a major impact in the development of deontic logic is Jørgensen's dilemma. In a nutshell, given that norms cannot be true or false, the dilemma implies that deontic logic cannot be based on traditional truth functional semantics. In particular, building on a tradition of Alchourrón and Bulygin in the seventies, Makinson [84] argues that norms need to be represented explicitly. SDL, DSDL and STIT logic represent logical relations between deontic operators, but they do not explicitly represent a distinction between norms and obligations. The explicit representation of norms is the basis of alternative semantics, that breaks with the idea of traditional semantics that norms and obligations have truth values, and most importantly, that discards the main technical and conceptual tool of traditional semantics, namely possible worlds. As an example, in this section we illustrate this alternative semantics using input/output logic.

5.1 Jørgensen's dilemma

While normative concepts are the subject of deontic logic, it is quite difficult to see how there can be a logic of such concepts at all. Norms like individual imperatives, promises, legal statutes, and moral standards are usually not viewed as being true or false. E.g. consider imperative or permissive expressions such as "John, leave the room!" and "Mary, you may enter now": they do not describe, but demand or allow a behavior on the part of John and Mary. Being non-descriptive, they cannot meaningfully be termed true or false. Lacking truth values, these expressions cannot—in the usual sense—be premise or conclusion in an inference, be termed consistent or contradictory, or be compounded by truth-functional operators. Hence, though there certainly exists a logical study of normative expressions and concepts, it seems there cannot be a logic of norms: this is Jørgensen's dilemma [65, 84].

Though norms are neither true nor false, one may state that *according to the norms*, something ought to be done or is permitted: the statements "John ought to leave the room" and"Mary is permitted to enter" are then true or false descriptions of the normative situation. Such statements are sometimes called normative statements, as distinguished from norms. To express principles such as the principle of conjunction: $O(p \wedge q) \leftrightarrow (Op \wedge Oq)$, with Boolean operators having truth-functional meaning at all places, deontic logic has resorted to interpreting its formulas Op, Fp, Pp not as representing norms, but as representing such normative statements. A possible logic of normative statements may then reflect logical properties of underlying norms—thus logic may have a "wider reach than truth", as Von Wright [124] famously stated.

Since the truth of normative statements depends on a normative situation, in the way in which the truth of the statement "John ought to leave the room" depends on whether some authority ordered John to leave the room or not, it seems that norms must be represented in a logical semantics that models such truth or falsity. However, semantics used to model the truth or falsity of normative statements mostly fail to include norms. Standard deontic semantics evaluates deontic formulas with respect to sets of worlds, in which some are ideal or better than others—Ox is then defined to be true if x is true in all ideal or the best reachable worlds. Alternatively, norms, not ideality, should provide the basis on which normative statements are evaluated. Thus the following question arises, asked by D. Makinson [84]:

Challenge 5. *How can deontic logic be reconstructed in accord with the philosophical position that norms are neither true nor false?*

In the older literature on deontic logic there has been a veritable 'imperativist tradition' of authors that have, deviating from the standard approach, in one way or other, tried to give truth definitions for deontic operators with respect to given sets of norms. Cf. among others S. Kanger [67], E. Stenius [105], T. J. Smiley [103], Z. Ziemba [125], B. van Fraassen [114], Alchourrón and Bulygin [2] and I. Niiniluoto [90]. The reconstruction of deontic logic as logic about imperatives has been the project of Jörg Hansen beginning with [47]. Input/output logic [85] is another reconstruction of a logic of norms in accord with the philosophical position that norms direct rather than describe, and are neither true nor false. We explain it in more detail in the next section below.

5.2 Input/output logic

To illustrate a possible answer to the dilemma, we use Makinson and van der Torre's input/output logic [85, 86, 87], and we therefore assume familiarity with this approach (cf. [88] for an introduction). Input/output logic takes a very general view at the process used to obtain conclusions (more generally: outputs) from given sets of premises (more generally: inputs). While the transformation may work in the usual way, as an 'inference motor' to provide logical conclusions from a given set of premises, it might also be put to other, perhaps non-logical uses. Logic then acts as a kind of secretarial assistant, helping to prepare the inputs before they go into the machine, unpacking outputs as they emerge, and, less obviously, coordinating the two. The process as a whole is one of logically assisted transformation, and is an inference only when the central transformation is so. This is the general perspective underlying input/output logic. It is one of logic at work rather than logic in isolation; not some kind of non-classical logic, but a way of using the classical one.

Suppose that we have a set G (meant to be a set of conditional norms), and a set A of formulas (meant to be a set of given facts). The problem is then: how may we reasonably define the set of propositions x making up the output of G given A, which we write $out(G, A)$? In particular, if we view the output as a collection of descriptions of states of affairs that ought to obtain given the norms G and the facts A, what is a reasonable output operation that enables us to define a deontic O-operator that returns the normative statements that are true given the norms and the facts—the normative consequences given the situation? One such definition is the following:

$$G, A \models Ox \quad \text{iff} \quad x \in out(G, A)$$

So Ox is true iff the output of G under A includes x. Note that this is rather a description of how we think such an output should or might be interpreted, whereas 'pure' input/output logic does not discuss such definitions. For a simple case, let G include a conditional norm that states that if a is the case, x should obtain (we write $(a, x) \in G$). An unconditional norm that commits the agent to realizing x is represented by a conditional norm $(\top, x)$, where $\top$ means an arbitrary tautology. If a can be inferred from A, i.e. if $a \in Cn(A)$, and z is logically implied by x, then z should be among the normative consequences of G given A. An operation that does this is simple-minded output out_1:

$$out_1(G, A) \quad = \quad Cn(G(Cn(A)))$$

where $G(B) = \{y \mid (b, y) \in G \text{ and } b \in B\}$. So in the given example, Oz is true given $(a, x) \in G$, $a \in Cn(A)$ and $z \in Cn(x)$.

Simple-minded output may, however, not be strong enough. Sometimes, legal argumentation supports reasoning by cases: if there is a conditional norm (a, x) that states that an agent must bring about x if a is the case, and a norm (b, x) that states that the same agent must also bring about x if b is the case, and $a \lor b$ is implied by the facts, then we should be able to conclude that the agent must bring about x. An operation that supports such reasoning is basic output out_2:

$$out_2(G, A) \quad = \quad \cap \{Cn(G(V)) \mid v(A) = 1\}$$

where v ranges over Boolean valuations plus the function that puts $v(b) = 1$ for all formulae b, and $V = \{b \mid v(b) = 1\}$. It can easily be seen that now Ox is true given $\{(a, x), (b, x)\} \subseteq G$ and $a \lor b \in Cn(A)$.

This definition of out_2 may give rise to a mere feeling of merely technical adequacy, because of its recourse to intersection and valuations, neither of which quite corresponds to our natural course of reasoning in such situations. However, this semantics makes explicit what is present but implicit in the use of possible worlds in conditional logics: if you want to reason by cases in the logic, you need to represent the cases explicitly in the semantics.

It is quite controversial whether reasoning with conditional norms should support 'normative' or 'deontic detachment', i.e. whether it should be accepted that if one norm (a, x) commands an agent to make x true in conditions a, and another norm (x, y) directs the agent to make y true given x is true, then the agent has an obligation to make y true if a is factually true. Some would argue that as long as the agent has not in fact realized x, the norm to bring about y is not 'triggered'; others would maintain that obviously the agent has an obligation to make $x \wedge y$ true given that a is true. Moreover, the inference can be restricted to cases where the agent ought to make x true instantly rather than eventually, see [84, 11] If such detachment is viewed as permissible for normative reasoning, then one might use reusable output out_3 that supports such reasoning:

$$out_3(G, A) = \cap \{Cn(G(B)) \mid A \subseteq B = Cn(B) \supseteq G(B)\}$$

An operation that combines reasoning by cases with deontic detachment is then reusable basic output out_4:

$$out_4(G, A) = \cap \{Cn(G(V)) : v(A) = 1 \text{ and } G(V) \subseteq V\}$$

It may turn out that further modifications of the output operation are required in order to produce reasonable results for normative reasoning. Also, the proposal to employ input/output logic to reconstruct deontic logic may lead to competing solutions, depending on what philosophical views as to what transformations should be acceptable one subscribes to. All this is what input/output logic is about. However, it should be noted that input/output logic succeeds in representing norms as entities that are neither true nor false, while still permitting normative reasoning about such entities.

5.3 Contrary to duty reasoning reconsidered

In the input/output logic framework, the strategy for eliminating excess output is to cut back the set of generators to just below the threshold of yielding excess. To do that, input/output logic looks at the maximal non-excessive subsets, as described by the following definition:

Definition (Maxfamilies) *Let G be a set of conditional norms and A and C two sets of propositional formulas. Then $maxfamily(G, A, C)$ is the set of maximal subsets $H \subseteq G$ such that $out(H, A) \cup C$ is consistent.*

For a possible solution to Chisholm's paradox, consider the following output operation $out^\cap$:

$$out^\cap(G, A) = \cap \{out(H, A) \mid H \in maxfamily(G, A, A)\}$$

So an output x is in $out^\cap(G, A)$ if it is in output $out(H, A)$ of all maximal norm subsets $H \subseteq G$ such that $out(H, A)$ is consistent with the input A. Let a deontic O-operator be defined in the usual way with regard to this output:

$$G, A \models O^\cap x \quad \text{iff} \quad x \in out^\cap(G, A)$$

Furthermore, tentatively, and only for the task of shedding light on Chisholm's paradox, let us define an entailment relation between norms as follows:

Definition (Entailment relation) *Let G be a set of conditional norms, and (a, x) be a norm whose addition to G is under consideration. Then (a, x) is entailed by G iff for all sets of propositions A, $out^{\cap}(G \cup \{(a, x)\}, A) = out^{\cap}(G, A)$.*

So a (considered) norm is entailed by a (given) set of norms if its addition to this set would not make a difference for any set of facts A. Finally, let us use the following cautious definition of 'coherence from the start' (also called 'minimal coherence' or 'coherence per se'), see Section 7:

A set of norms G is 'coherent from the start'   iff $\bot \notin out(G, \top)$.

Now consider a 'Chisholm norm set' $G = \{(\top, x), (x, z), (\neg x, \neg z), \}$, where $(\top, x)$ means the norm that the man must go to the assistance of his neighbors, (x, z) means the norm that it ought to be that if he goes he ought to tell them he is coming, and $(\neg x, \neg z)$ means the norm that if he does not go he ought not to tell them he is coming. It can be easily verified that the norm set G is 'coherent from the start' for all standard output operations out_n, since for these either $out(G, \top) = Cn(\{x\})$ or $out(G, \top) = Cn(\{x, z\})$, and both sets $\{x\}$ and $\{x, z\}$ are consistent. Furthermore, it should be noted that all norms in the norm set G are independent from each other, in the sense that no norm $(a, x) \in G$ is entailed by $G \setminus \{(a, x)\}$ for any standard output operation $out_n^{(+)}$: for $(\top, x)$ we have $x \in out^{\cap}(G, \top)$ but $x \notin out^{\cap}(G \setminus \{(\top, x)\}, \top)$, for (x, z) we have $z \in out^{\cap}(G, x)$ but $z \notin out^{\cap}(G \setminus \{(x, z)\}, x)$, and for $(\neg x, \neg z)$ we have $\neg z \in out^{\cap}(G, \neg x)$ but $\neg z \notin out^{\cap}(G \setminus \{(\neg x, \neg z)\}, \top)$. Finally consider the 'Chisholm fact set' $A = \{\neg x\}$, that includes as an assumed unalterable fact the proposition $\neg x$, that the man will not go to the assistance of his neighbors: we have $maxfamily(G, A, A) = \{G \setminus \{(\top, x)\}\} = \{\{(x, z), (\neg x, \neg z), \}\}$ and either $out(G \setminus \{(\top, x)\}, A) = Cn(\{\neg z\})$ or $out(G \setminus \{(\top, x)\}, A) = Cn(\{\neg x, \neg z\})$ for all standard output operations $out_n^{(+)}$, and so $O^{\cap} \neg z$ is true given the norm and fact sets G and A, i.e. the man must not tell his neighbors he is coming. Thus:

$G, A \models O^{\cap} \neg z$

6 Multiagent detachment

In Section 6.1 we introduce normative multiagent systems using agents and controllable propositions, and we introduce a challenge for detachment for multiagent systems. In Section 6.2 we give a solution for the challenge in these formalisms.

6.1 Challenge for multiagent detachment

Olde Loohuis [91] argues that the assumption that other agents comply with their norms reflects that agents live in a responsible world. However, Makinson [84] observes that if all we know is that "John owes Peter $1000" and "if John pays Peter $1000, then Peter is obliged to give John a receipt," then we cannot detach that Peter has to give John a receipt unconditionally based on the assumption that John will pay Peter the money.

We assume that the normative system is known to all agents, and in this section we assume that it does not change over time, and that each norm is directed to one agent only. The agents reason about the consequences of the normative system, that is, which obligations and permissions can be detached from it. With an explicit normative system, the agents should act such that they do not violate norms. Moreover, in this section we assume that each (instance of a) norm specifies the behavior of a single individual agent. For example, a norm may say that an agent should drive to the right hand side of the street, but we do not consider group norms saying that agents should live together in harmony.

We do not assume a full action theory as in STIT logic, but we assume a minimal action theory: the set of propositions is partitioned into parameters (uncontrollable propositions) and decision variables (controllable propositions). Boutilier [19] traces this idea back to discrete event systems, see also Cholvy and Garion [30]. It is an abstract and general approach, since we can instantiate the propositions with action descriptions like do(action) or done(action). Note that this generality is in line with game theory, which abstracts away sequential decisions in extensive games by representing conditional plans as strategic games. Boutilier observes that the theory can be extended to a full fledged action theory by, for example, introducing a causal theory. By convention, the proposition letters p, p_1, etc are parameters, a, a_1, ..., are decision variables for agent 1, b, b_1, ..., are decision variables for agent 2, etc. Norms are written as pairs of propositional formulas, where (p_1, p_2) is read as "if p_1 is the case, then p_2 ought to be the case," (a_1, a_2) is read as "if agent 1 does a_1, then he has to do a_2," and so on. We restrict the propositional language to conjunctions of literals (propositional atoms or their negations), so we do not consider disjunctions or material implications.

Definition 6.1 (Normative multi agent system, individual norms). *A normative multiagent system is a tuple NMAS= $\langle A, P, c, N \rangle$ where A is a set of agents, P is a set of atomic propositions, $c : P \to A$ is a partial function which maps the propositions to the agents controlling them, and N is a set of pairs of conjunctions of literals built of P, such that if $(\phi, \psi) \in N$, then all propositional atoms in ψ are controlled by a single agent.*

Our action theory may be seen as a simple kind of STIT theory, in the sense that an obligation for a proposition p controlled by agent α may be read as: "the agent α ought to see to it that p is the case." Though this abstracts away from the temporal issues of STIT operators, it still has the characteristic property of STIT logics that actions have a higher granularity than worlds.

Makinson [84] illustrates the intricacies of temporal reasoning with norms, obligations and agents by discussing the iteration of detachment, in the sense that from the two conditional norms "if ϕ, then obligatory ψ" and "if ψ, then obligatory χ" together with the fact ϕ, we can derive not only that ψ is obligatory, but also that χ is obligatory. Makinson's challenge is how to detach obligations based on the principle that agents cannot assume that other agents comply with their norms, but they assume that they themselves comply with their norms. In other words, deontic detachment holds only for the single agent a-temporal case.

First, Makinson argues that iteration of detachment often appears to be appropriate. He gives the following example, based on instructions to authors preparing manuscripts.

Example 6.2 (Manuscript [84]). *Let the set of norms be as follows: $(25x15, 12)=$ "if $25x15$, then obligatory 12" and $(12, refs10)=$ "if 12, then obligatory $refs10$", where $25x15$ is "The text area is 25 by 15 cm", 12 is "The font size for the main text is 12 points", and $refs10$ is "The font size for the list of references is 10 points". Moreover, consider a single agent controlling the three variables. If the facts contain $25x15$, then we want to detach not only that it is obligatory that 12, but also that it is obligatory that $refs10$.*

Second, he argues that iteration of detachment sometimes appears to be inappropriate by discussing the following example, which he attributes to Sven Ove Hansson.

Example 6.3 (Receipt [84]). *Let instances of the norms be*
 $(owe_{jp}, pay_{jp})=$ "if owe_{jp}, then obligatory pay_{jp}" and
 $(pay_{jp}, receipt_{pj})=$ "if pay_{xy}, then obligatory $receipt_{pj}$"
where owe_{jp} is "John owes Peter \$1000", pay_{jp} is "John pays Peter \$1000", and $receipt_{pj}$ is "Peter gives John a receipt for \$1000". Moreover, assume that the first variable is not controlled by an agent, the second is controlled by John, and the third is controlled by Peter. Intuitively Makinson would say that in the circumstance that John owes Peter \$1000, considered alone, Peter has no obligation to write any receipt. That obligation arises only when John fulfils his obligation.

Makinson observes that there appear to be two principal sources of difficulty here. One concerns the passage of time, and the other concerns bearers of the obligations. Sven Ove Hansson's example above involves both of these factors.

> "We recall that our representation of norms abstracts entirely from the question of time. Evidently, this is a major limitation of scope, and leads

to discrepancies with real-life examples, where there is almost always an implicit time element. This may be transitive, as when we say "when b holds then a should eventually hold", or "... should simultaneously hold". But it may be intransitive, as when we say "when b holds then a should hold within a short time" or "... should be treated as a matter of first priority to bring about". Clearly, iteration of detachment can be legitimate only when the implicit time element is either nil or transitive. Our representation also abstracts from the question of bearer, that is, who (if anyone) is assigned responsibility for carrying out what is required. This too can lead to discrepancies. Iteration of detachment becomes questionable as soon as some promulgations have different bearers from others, or some are impersonal (i.e. without bearer) while others are not. Only when the locus of responsibility is held constant can such an operation take place." [84]

Challenge 6. *How to define detachment for multiple agents?*

Broersen and van der Torre [21] consider the temporal aspects of the example. In this section we consider the actions of the agents. The following example extends the discussion of the example to aggregative deontic detachment.

Example 6.4 (continued). *Consider again* (owe_{jp}, pay_{jp}) *and* ($pay_{jp}, receipt_{pj}$), *where the first variable is not controlled by an agent, the second is controlled by John, and the third is controlled by Peter. In the circumstance that John owes Peter $1000, considered alone, do we want to derive the obligation for $pay_{jp} \wedge receipt_{pj}$, that is, the obligation that "John pays Peter $1000", and "Peter gives John a receipt for $1000"? In many systems the obligation for $pay_{jp} \wedge receipt_{pj}$ implies the obligation for $receipt_{pj}$, such that the answer will be negative. However, if the obligation for $pay_{jp} \wedge receipt_{pj}$ does not imply the obligation for $receipt_{pj}$, then maybe the obligation for $pay_{jp} \wedge receipt_{pj}$ is not as problematic as the obligation for $receipt_{pj}$. Moreover, the obligation for $pay_{jp} \wedge receipt_{pj}$ is a compact representation of the fact that ideally, the exchange of money and receipt takes place.*

6.2 Deontic detachment for agents

As the iterative approaches seem most natural to most people, we define deontic detachment of agents using these iterative approaches. The question thus arises whether we consider sequential or iterated detachment. The following example illustrates this question, not discussed by Makinson [84].

Example 6.5. $N = \{(p,a), (a, b_1), (a \wedge b_1, b_2)\}$ where p is a parameter, a is a decision variable of agent 1, and b_1 and b_2 are decision variables of agent 2. In context $F = \{p, a\}$, do we want to detach only b_1, or both b_1 and b_2? If we can detach b_2, then this implies that despite the fact that a and b_1 are decision variable from distinct agents we can use $(a \wedge b_1, b_2)$ to detach b_2.

In the above example, we believe that b_2 should be derivable, because only b_1 is reused when b_2 is detached, and both b_1 and b_2 are decision variables of the same agent. In other words, when considering the norm $(a \wedge b_1, b_2)$ to detach b_2, we should not consider the norm and reject it because there is a variable in the input which refers to another agent, but we should consider it since we have $a \in F$ as a fact, and b_1 already in the output, we can derive b_2 too.

If b_2 should not be derivable, then we could simply restrict the set of norms that we select from N to satisfy the syntactic criterion, just like we selected the set of norms N_0. However, if b_2 should be derivable, then we have to define detachment procedures for each agent, and combine them afterwards. This is formalized in the following detachment procedure for agents.

Definition 6.6 (Iterative detachment for agents.). *Agent $a \in A$ controls a propositional formula ϕ, written as $c(\phi) = a$, if and only if for all atoms $x \in \phi$ we have $c(x) = a$.*

$$N_0^a = \{(\phi, \psi) \in N \mid F \cup \{\phi\} \not\models \neg\psi, c(\psi) = a\}$$

$E_0^{ia} = \emptyset$. *For $n = 1$ to ∞ do $E_{n+1}^{ia} = \{\psi \mid (\phi, \psi) \in N_0^a, F \cup E_n^{ia} \models \phi\}$ if consistent with F, E_n^{ia} otherwise. $out^{ia}(N, F, a) = Cn(\cup E_i^{ia})$, and $out^{ia}(N, F) = \cup_{a \in A} out^{ia}(N, F, a)$.*

We leave the logical analysis of this ans related approaches to future work.

7 Coherence

Consider norms which at the same time require you to leave the room and not to leave the room. In such cases, we are inclined to say that there is something wrong with the normative system. This intuition is captured by the SDL axiom $D : \neg(Ox \wedge O\neg x)$ that states that there cannot be co-existing obligations to bring about x and to bring about $\neg x$, or, using the standard cross-definitions of the deontic modalities: x cannot be both, obligatory and forbidden, or: if x is obligatory then it is also permitted. However, what does this tell us about the normative system?

Since norms do not bear truth values, we cannot, in any usual sense, say that such a set of norms is inconsistent. All we can consider is the consistency of the output of a set of norms. We like to use the term *coherence* with respect to a set of norms with consistent output. For a start, consider the notion of minimal coherence in Section 5.3:

(0) A set of norms G is minimal coherent   iff $\bot \notin out(G, \emptyset)$.

This is clearly very weak, as for example the norms $(a, x), (a, \neg x)$ would be coherent. Alternatively, we might try to define coherence as follows:

(1) A set of norms G is coherent   iff $\bot \notin out(G, A)$.

However, this definition seems not quite sufficient: one might argue that one should be able to determine whether a set of norms G is coherent or not regardless of what arbitrary facts A might be assumed. A better definition would be (1a):

(1a) A set of norms G is coherent   iff   there exists a set of formulas A such that $\bot \notin out(G, A)$.

For (1a) it suffices that there exists a situation in which the norms can be, or could have been, fulfilled. However, consider the set of norms $G = \{(a, x), (a, \neg x)\}$ that requires both x to be realized and $\neg x$ to be realized in conditions a: it is immediate that e.g. for all output operations out_n, we have $\bot \notin out_n(G, \neg a)$: no conflicting demands arise when $\neg a$ is factually assumed. Yet something seems wrong with a normative system that explicitly considers a fact a only to tie to it conflicting normative consequences. The dual of (1a) would be

(1b) A set of norms G is coherent   iff   for all sets of formulas A, we have $\bot \notin out(G, A)$.

Now a set G with $G = \{(a, x), (a, \neg x)\}$ would no longer be termed coherent. (1b) makes the claim that for no situation A, two norms $(a, x), (b, y)$ would ever come into conflict, which might seem too strong. We may wish to restrict A to sets of facts that are consistent, or that are not in violation of the norms. The question is, basically, how to distinguish situations that the norm-givers should have taken care of, from those that describe misfortune or otherwise unhappy circumstances. A weaker claim than (1b) would be (1c):

(1c) A set of norms G is coherent   iff   for all a with $(a, x) \in G$, we have $\bot \notin out(G, a)$.

By this change, consistency of output is required just for those factual situations that the norm-givers have foreseen, in the sense that they have explicitly tied normative consequences to such facts. Still, (1c) might require further modification, since if a is a foreseen situation, and so is b, then also $a \vee b$ or $a \wedge b$ might be counted as foreseen situations for which the norms should be coherent.

As one anonymous reviewer suggested, another solution consists in combining elements of previous proposals:

(1d)   A set of norms G is coherent   iff   for each $A \subseteq \{a \mid (a,x) \in G\}$, if A is non-empty and consistent, then $\bot \notin out(G, A)$.

However, there is a further difficulty: let G contain a norm $(a, \neg a)$ that, for conditions in which a is unalterably true, demands that $\neg a$ be realized. We then have $\neg a \in out_n(G, a)$ for the principal output operations out_n, but not $\bot \in out_n(G, a)$. Certainly the term 'incoherent' should apply to a normative system that requires the agent to accomplish what is—given the facts in which the duty arises—impossible. However, since not every output operation supports 'throughput', i.e. the input is not necessarily included in the output, neither (1) nor its variants implies that the agent can actually realize all propositions in the output, though they might be logically consistent. We might therefore demand that the output be not merely consistent, but consistent with the input:

(2)   A set of norms G is coherent   iff $out(G, A) \cup A \not\models \bot$.

However, with definition (2) we obtain the questionable result that for any case of norm-violation, i.e. for any case in which $(a, x) \in G$ and $(a \wedge \neg x) \in Cn(A)$, G must be termed incoherent—Adam's fall would only indicate that there was something wrong with God's commands. One remedy would be to leave aside all those norms whose violation is entailed by the circumstances A, i.e. instead of $out(G, A)$ consider $out(\{(a, x) \in G \mid (a \wedge \neg x) \notin Cn(A)\}, A)$—but then a set G such that $(a, \neg a) \in G$ would not be incoherent.[2] It seems it is time to formally state our problem:

Challenge 7. *When is a set of norms to be termed 'coherent'?*

As can be seen from the discussion above, input/output logic provides the tools to formally discuss this question, by rephrasing the question of coherence of the norms as one of consistency of output, and of output with input. Both notions have been explored in the input/output framework as 'output under constraints', see also the motivation regarding contrary-to-duty reasoning in Section 1.4.:

Definition (Output under constraints) *Let G be a set of conditional norms and A and C two sets of propositional formulas. Then G is coherent in A under constraints C when $out(G, A) \cup C$ is consistent.*

Future study must define an output operation, determine the relevant states A, and find the constraints C, such that any set of norms G would be appropriately termed coherent or incoherent by this definition.

---

[2]Temporal dimensions are not considered here. In an approach that would consider dynamic norms, one may argue, throughput should not be included in a definition of coherence as any change involves an inconsistency between the way things were and the way they become.

8 Normative conflicts and dilemmas

There are essentially two views on the question of normative conflicts: in the one view, they do not exist. In the other view, conflicts and dilemmas are ubiquitous.

According to the view that normative conflicts are ubiquitous, it is obvious that we may become the addressees of conflicting normative demands at any time. My mother may want me to stay inside while my brother wants me to go outside with him and play games. I may have promised to finish a paper by the end of a certain day, while for the same day I have promised a friend to come to dinner—now it is late afternoon and I realize I will not be able to finish the paper if I visit my friend. Social convention may require me to offer you a cigarette when I am lighting one for myself, while concerns for your health should make me not offer you one. Legal obligations might collide - think of the case where the SWIFT international money transfer program was required by US anti-terror laws to disclose certain information about its customers, while under European law that also applied to that company, it was required not to disclose this information. Formally, let there be two conditional norms (a, x) and (b, y): unless we have that either $(x \to y) \in Cn(a \wedge b)$ or $(y \to x) \in Cn(a \wedge b)$ there is a possible situation $a \wedge b \wedge \neg(x \wedge y)$ in which the agent can still satisfy each norm individually, but not both norms collectively. But to assume this for any two norms (a, x) and (b, y) is clearly absurd. Nevertheless, as discussed extensively in Section 1 of this article, Lewis's [74, 75] and Hansson's [53] deontic semantics imply that there exists a 'system of spheres', in our setting: a sequence of boxed contrary-to-duty norms $(\top, x_1), (\neg x_1, x_2), (\neg x_1 \wedge \neg x_2, x_3), \ldots$ that satisfies this condition. So any logic about norms must take into account possible conflicts. But standard deontic logic SDL includes D: $\neg(Ox \wedge O\neg x)$ as one of its axioms, and it is not immediately clear how deontic reasoning could accommodate conflicting norms.

Challenge 8a. *How can deontic logic accommodate possible conflicts of norms?*

The literature on normative conflicts and dilemmas is vast. As highlighted earlier in this article, here we do not aim at an exhausting literature review on the topic; for that, the interested reader is referred to Goble's [38] chapter in the handbook of deontic logic and normative systems. If we accept the view that normative conflicts not only genuinely exist but are also ubiquitous, one classical way to deal with such conflicts consists in denying that 'ought' implies 'can', as done by Lemmon [73]. Another common solution is to deny the principle of conjunction, that is, to deny that oughting to do x and y separately implies ought to do both [89, 114, 35]. However, this solution was challenged by Horty's example [59, 60, 61, 62] where, from "Smith ought to fight in the army or perform alternative national service" and "Smith ought not to fight in the army", we should be able to derive "Smith ought to perform alternative national service". By withdrawing the principle of conjunction, this argument is no longer valid. The distribution rule states that x necessitates y implies that, if one ought to do x, then one ought to do y. As Goble [38] observes, although this principle has been often criticized for its role in many deontic paradoxes, its responsibility in connection with normative conflicts has rarely been discussed. Keeping the principle of conjunction while removing the distribution rule would validate Horty's argument [37]. For other systems that restrict the distribution principle, see [36, 37].

In an input/output setting one could say that there exists a conflict whenever $\bot \in Cn(out(G, A) \cup A)$, i.e. whenever the output is inconsistent with the input: then the norms cannot all be satisfied in the given situation. There appear to be two ways to proceed when such inconsistencies cannot be ruled out. For the concepts underlying the 'some-things-considered' and 'all-things-considered' O-operators defined below cf. Horty [60] and Hansen [48, 49]. For both, it is necessary to recur to the the notion of a $maxfamily(G, A, A)$, i.e. the family of all maximal $H \subseteq G$ such that $out(H, A) \cup A$ is consistent. On this basis, input/output logic defines the following two output operations $out^\cup$ and $out^\cap$:

$$out^\cup(G, A) = \bigcup \{out(H, A) \mid H \in maxfamily(G, A, A)\}$$
$$out^\cap(G, A) = \bigcap \{out(H, A) \mid H \in maxfamily(G, A, A)\}$$

Note that $out^\cup$ is a non-standard output operation that is not closed under consequences, i.e. we do not generally have $Cn(out^\cup(G, A)) = out^\cup(G, A)$. Finally we may use the intended definition of an O-operator

$$G, A \models Ox \quad \text{iff} \quad x \in out(G, A)$$

to refer to the operations $out^\cup$ and $out^\cap$, rather than the underlying operation $out(G, A)$ itself, and write $O^\cup x$ and $O^\cap x$ to mean that $x \in out^\cup(G, A)$ and $x \in out^\cap(G, A)$, respectively. Then we have that the 'some-things-considered', or 'bold' O-operator $O^\cup$ describes x as obligatory given the set of norms G and the facts A if x is in the output of some $H \in maxfamily(G, A, A)$, i.e. if some subset of non-conflicting norms, or: some coherent normative standard embedded in the norms, requires x to be true. It is immediate that neither the SDL axiom $D : \neg(Ox \wedge O\neg x)$ nor the agglomeration principle $C : Ox \wedge Oy \to O(x \wedge y)$ holds for $O^\cup$, as there may be two competing standards demanding x and $\neg x$ to be realized, while there may be none that demands the impossible $x \wedge \neg x$. However, the 'all-things-considered', or 'sceptic', O-operator $O^\cap$ describes x as obligatory given the norms G and the facts A if x is in the outputs of all $H \in maxfamily(G, A, A)$, i.e. it requires that x must be realized according to all coherent normative standards. Note that by this definition, both SDL theorems D and C are validated.

The opposite view, that normative conflicts do not exist, appeals to the very notion of obligation: it is essential for the function of norms—to direct human behavior—that the subject of the norms is capable of following them. To state a norm that cannot be fulfilled is a meaningless use of language. To state two norms which cannot both be fulfilled is confusing the subject, not giving him or her directions. To say that a subject has two conflicting obligations is therefore a misuse of the term 'obligation'. So there cannot be conflicting obligations, and if things appear differently, a careful inspection of the normative situation is required that resolves the dilemma in favor of the one or other of what only appeared both to be obligations. In particular, this inspection may reveal that the apparent conflicts in reality comes from some ambiguities in the examples, for instance where a moral 'ought' is not compatible with a legal 'ought': thus, there is no real conflict, because the two 'oughts' refer to two different spheres, and each should be represented with a different operator [26, 27]. Or again, a priority ordering of the apparent obligations may help resolving the conflict, e.g. in Ross [100], von Wright [121, 122], and Hare [55]. The problem that arises for such a view is then how to determine the 'actual obligations' in face of apparent conflicts, or, put differently, in the face of conflicting 'prima facie' obligations.

Challenge 8b. How can the resolution of apparent conflicts be semantically modeled?

Again, both the $O^\cup$ and the $O^\cap$-operator may help to formulate and solve the problem: $O^\cup$ names the conflicting *prima facie* obligations that arise from a set of norms G in a given situation A, whereas $O^\cap$ resolves the conflict by only telling the agent to do what is required by all maximal coherent subsets of the norms: so there might be conflicting 'prima facie' $O^\cup$-obligations, but no conflicting 'all things considered' $O^\cap$-obligations. The view that a priority ordering helps to resolve conflicts seems more difficult to model. A good approach appears to be to let the priorities help us to select a set $P(G, A, A)$ of preferred maximal subsets $H \in$ *maxfamily*(G, A, A). We may then define the $O^\cap$-operator not with respect to the whole of *maxfamily*(G, A, A), but only with respect to its selected preferred subsets $P(G, A, A)$. Ideally, in order to resolve all conflicts, the priority ordering should narrow down the selected sets to $card(P(G, A, A)) = 1$, but this generally requires a strict ordering of the norms in G. The demand that all norms can be strictly ordered is itself subject of philosophical dispute. Some moral requirements may be incomparable: this is Sartre's paradox, where the requirement that Sartre's student stays with his ailing mother conflicts with the requirement that the student joins the resistance against the German occupation [101]. Other moral requirements may be of equal weight, e.g. two simultaneously obtained obligations towards identical twins, of which only one can be fulfilled [89]. The difficult part is then to define a mechanism that determines the preferred maximal subsets by use of the given priorities between the norms. There have been several proposals to this effect, not all of them successful, and the reader is referred to the discussions in Boella and van der Torre [13] and Hansen [50, 51].

9 Descriptive dyadic obligations

Dyadic deontic operators, that formalize e.g. 'x ought to be true under conditions a' as $O(x|a)$, were introduced over 50 years ago by G. H. von Wright [118]. Their introduction was due to Prior's paradox of derived obligation: often a primary obligation Ox is accompanied by a secondary, 'contrary-to-duty' obligation that pronounces y (a sanction, a remedy) as obligatory if the primary obligation is violated. At the time, the usual formalization of the secondary obligation would have been $O(\neg x \to y)$, but given Ox and the axioms of standard deontic logic SDL, $O(\neg x \to y)$ is derivable for any y. A bit later, Chisholm's paradox showed that formalizing the secondary obligation as $\neg x \to Oy$ produces similarly counterintuitive results. So to deal with such contrary-to-duty conditions, the dyadic deontic operator $O(x|a)$ was invented. For a historical account the reader is referred to Hilpinen and McNamara's chapter in the handbook of deontic logic and normative systems [57].

In Section 1.3 we have extensively discussed DSDL. The perhaps best-known semantic characterization of dyadic deontic logic is B. Hansson's [53] system DSDL3, axiomatized by Spohn [104]. Hansson's idea was that the circumstances (the conditions a) are something which has actually happened (or will unavoidably happen) and which cannot be changed afterwards. Ideal worlds in which $\neg a$ is true are therefore excluded. However, some worlds may still be better than others, and there should then be an obligation to make 'the best out of the sad circumstances". Consequently, Hansson presents a possible worlds semantics in which all worlds are ordered by a preference (betterness) relation. $O(x|a)$ is then defined true if x is true in the best a-worlds. Here, we intend to employ semantics that do not make use of any prohairetic betterness relation, but that model deontic operators with regard to given sets of norms and facts.

Challenge 9. *How to define dyadic deontic operators with regard to given sets of norms and facts?*

Input/output logic assumes a set of (conditional) norms G, and a set of unalterable facts A. The facts A may describe a situation that is inconsistent with the output $out(G, A)$: suppose there is a primary norm $(\top, a) \in G$ and a secondary norm $(\neg a, x) \in G$, i.e. $G = \{(\top, a), (\neg a, x)\}$, and $A = \{\neg a\}$. Though $a \in out(G, A)$, it makes no sense to describe a as obligatory since a cannot be realized any more in the given situation—no crying over spilt milk. Rather, the output should include only the consequent of the secondary obligation x—it is the best we can make out of these circumstances. To do so, we return to the definitions of $\mathit{maxfamily}(G, A, A)$ as the set of all maximal subsets $H \subseteq G$ such that $out(H, A) \cup A$ is consistent, and the set $out^\cap(G, A)$ as the intersection of all outputs from $H \in \mathit{maxfamily}(G, A, A)$, i.e. $out^\cap(G, A) = \bigcap \{out(H, A) \mid H \in \mathit{maxfamily}(G, A, A)\}$. We may then define:

$$G \models O(x|a) \quad \text{iff} \quad x \in out^\cap(G, \{a\})$$

Thus, relative to the set of norms G, $O(x|a)$ is defined true if x is in the output under a of all maximal sets H of norms such that their output under $\{a\}$ is consistent with a. In the example where $G = \{(\top, a), (\neg a, x)\}$ we therefore obtain $O(x|\neg a)$ but not $O(a|\neg a)$ as being true, i.e. only the consequent of the secondary obligation is described as obligatory in conditions $\neg a$.

In the above definition, the antecedent a of the dyadic formula $O(x|a)$ makes the inputs explicit: the truth definition does not make use of any facts other than a. This may be unwanted; one might consider an input set A of *given* facts, and employ the antecedent a only to denote an additional, *assumed* fact. Still, the output should contradict neither the given nor the assumed facts, and the output should include also the normative consequences x of a norm (a, x) given the assumed fact a. This may be realized by the following definition:

$$G, A \models O(x|a) \quad \text{iff} \quad x \in out^\cap(G, A \cup \{a\})$$

So, relative to a set of norms G and a set of facts A, $O(x, a)$ is defined true if x is in the output under $A \cup \{a\}$ of all maximal sets H of norms such that their output under $A \cup \{a\}$ is consistent with $A \cup \{a\}$.

Hansson's description of dyadic deontic operators as describing defeasible obligations that are subject to change when more specific, namely contrary-to-duty situations emerge, may be the most prominent view, but it is by no means the only one. Earlier authors like von Wright [119, 120] and Anderson [5] have proposed more normal conditionals, which in particular support 'strengthening of the antecedent' SA $O(x|a) \to O(x|a \wedge b)$. From an input/output perspective, such operators can be accommodated by defining

$$G, A \models O(x|a) \quad \text{iff} \quad x \in out(G, A \cup \{a\})$$

It is immediate that for all standard output operations out_n this definition validates SA. The properties of dyadic deontic operators that are, like the above, semantically defined within the framework of input/output logic, have not been studied so far. The theorems they validate will inevitably depend on what output operation is chosen, cf. Hansen [51] for some related conjectures.

10 Permissive norms

In formal deontic logic, permission is studied less frequently than obligation. For a long time, it was naively assumed that it can simply be taken as a dual of obligation, just as possibility is the dual of necessity in modal logic. Permission is then defined as the absence of an obligation to the contrary, and the modal operator P defined by $Px =_{def} \neg O \neg x$. Today's focus on obligations is not only in stark contrast how deontic logic began, for when von Wright [117] started modern deontic logic in 1951, it was the P-operator that he took as primitive, and defined obligation as an absence of a permission to the contrary. Rather, more and more authors have come to realize how subtle and multi-faceted the concept of permission is. Much energy was devoted to solving the problem of 'free choice permission', where one may derive from the statement that one is permitted to have a cup of tea or a cup of coffee that it is permitted to have a cup of tea, and it is permitted to have a cup of coffee, or for short, that $P(x \vee y)$ implies Px and Py (cf. Kamp [66]). Von Wright, in his late work starting with [123], dropped the concept of inter-definability of obligations and permissions altogether by introducing P-norms and O-norms, where one may call something permitted only if it derives from the collective contents of some O-norms and at most one P-norm. This concept of 'strong permission' introduced deontic 'gaps': whereas in standard deontic logic SDL, $O \neg x \vee Px$ is a tautology, meaning that any state of affairs is either forbidden or permitted, von Wright's new theory means that in the absence of explicit P-norms only what is obligatory is permitted, and that nothing is permitted if also O-norms are missing. Perhaps most importantly, Bulygin [24] observed that an authoritative kind of permission must be used in the context of multiple authorities and updating normative systems: if a higher authority permits you to do something, a lower authority can no longer prohibit it. Summing up, the understanding of permission is still in a less satisfactory state than the understanding of obligation and prohibition. Indeed, a whole chapter in the handbook of deontic logic and normative systems is devoted to the various forms of permission [54].

Challenge 10. *How to distinguish various kinds of permissions and relate them to obligations?*

From the viewpoint of input/output logic, one may first try to define a concept of negative permission in the line of the classic approach. Such a definition is the following:

$$G, A \models P^{neg}x \quad \text{iff} \quad \neg x \notin out(G, A)$$

So something is permitted by a code iff its negation is not obligatory according to the code and in the given situation. As innocuous and standard as such a definition seems, questions arise as to what output operation *out* may be used. Simple-minded output out_1 and basic output out_2 produce counterintuitive results: consider a set of norms G of which one norm (*work, tax*) demands that if I am employed then I have to pay taxes. For the default situation $A = \{\top\}$ then $P^{neg}(work \wedge \neg tax)$ is true, i.e. it is by default permitted that I am employed and do not pay taxes. Stronger output operations out_3 and out_4 that warrant reusable output exclude this result, but their use in deontic reasoning is questionable due to contrary-to-duty reasoning, as discussed in Section 1.

In contrast to a concept of negative permission, one may also define a concept of 'strong' or 'positive permission'. This requires a set P of explicit permissive norms, just as G is a set of explicit obligations. As a first approximation, one may say that something is positively permitted by a code iff the code explicitly presents it as such. However, this leaves a central logical question unanswered as to how explicitly given permissive and obligating norms may generate permissions that—in some sense—follow from the explicitly given norms. Pursuing von Wright's later approach, we may define:

$$G, P \models P^{stat}(x/a) \quad \text{iff} \quad x \in out(G \cup \{(b,y)\}, a) \text{ for some } (b,y) \in P \cup \{(\top, \top)\}$$

So there is a permission to realize x in conditions a if x is generated under these conditions either by the norms in G alone, or the norms in G together with some explicit permission (b, y) in P. We call this a 'static' version of strong permission. For example, consider a set G consisting of the norm (*work, tax*), and a set P consisting of the sole license (*18y, vote*) that permits all adults to take part in political elections. Then all of the following are true: $P^{stat}(tax/work)$, $P^{stat}(vote/18y)$, $P^{stat}(tax/work \wedge male)$ and also $P^{stat}(vote/\neg work \wedge 18y)$ (so even unemployed adults are permitted to vote).

Where negative permission is liberal, in the sense that anything is permitted that does not conflict with one's obligations, the concept of static permission is quite strict, as nothing is permitted that does not explicitly occur in the norms. In between, one may define a concept of 'dynamic permission' that defines something as permitted in some situation a if forbidding it for these conditions would prevent an agent from making use of some explicit (static) permission. The formal definition reads:

$$G, P \models P^{dyn}(x/a) \quad \text{iff} \quad \neg y \in out(G \cup \{(a, \neg x)\}, b) \text{ for some } y \text{ and conditions } b \text{ such that } G, P \models P^{stat}(y/b)$$

Consider the above static permission $P^{stat}(vote/\neg work \wedge 18y)$ that even the unemployed adult populations is permitted to vote, generated by $P = \{(18y, vote)\}$ and $G = \{(work, tax)\}$. We might also like to say, without reference to age, that the unemployed are protected from being forbidden to vote, and in this sense are permitted to vote, but $P^{stat}(vote/\neg work)$ is not true. And we might like to say that adults are protected from being forbidden to vote unless they are employed, and in this sense are permitted to be both unemployed and take part in elections, but also $P^{stat}(\neg work \wedge vote/18y)$ is not true. Dynamic permissions allow us to express such protections, and make both $P^{dyn}(vote/\neg work)$ and $P^{dyn}(\neg work \wedge vote/18y)$ true: if either $(\neg work, \neg vote)$ or $(18y, (\neg work \to \neg vote))$ were added to G we would obtain $\neg vote$ as output in conditions $(\neg work \wedge 18y)$ in spite of the fact that, as we have seen, $G, P \models P^{stat}(vote/\neg work \wedge 18y)$.

The relation of permission and obligation can also be studied from a multi-agent perspective. Think of two brothers who are fighting for a toy, and the mother obliges the son who's playing with the toy to permit his brother to play as well.

There are, ultimately, a number of questions for all these concepts of permissions that Makinson and van der Torre have further explored [87]. Other kinds of permissions have been discussed from an input/output perspective in the literature, too, for example permissions as exceptions of obligations [13]. It seems input/output logic is able to help clarify the underlying concepts of permission better than traditional deontic semantics. One challenge is Governatori's paradox [39], containing a conditional norm whose body and head are permissions: "the collection of medical information is permitted provided that the collection of personal information is permitted."

11 Meaning postulates and intermediate concepts

To define a deontic operator of individual obligation seems straightforward if the norm in question is an individual command or act of promising. For example, if you are the addressee α of the following imperative sentence

(1) You, hand me that screwdriver, please.

and you consider the command valid, then what you ought to do is to hand the screwdriver in question to the person β uttering the request. In terms of input/output logic, let x be the proposition that α hands the screwdriver to β: with the set of norms $G = \{(\top, x)\}$, the set of facts $A = \{\top\}$, and the truth definition Ox iff $x \in out(A, G)$: then we obtain that Ox is true, i.e. it is true that it ought to be that α hands the screwdriver to β.

Norms that belong to a legal system are more complex, and thus more difficult to reason about. Consider, for example

(2) An act of theft is punished by a prison sentence not exceeding 5 years or a fine.

Things are again easy if you are a judge and you know that the accused in front of you has committed an act of theft—then you ought to hand out a verdict that commits the accused to pay a fine or to serve a prison sentence not exceeding 5 years. However, how does the judge arrive at the conclusion that an act of theft has been committed? 'Theft' is a legal term that is usually accompanied by a legal definition such as the following one:

(3) Someone commits an act of theft if that person has taken a movable object from the possession of another person into his own possession with the intention to own it, and if the act occurred without the consent of the other person or some other legal authorization.

It is noteworthy that (3) is not a norm in the strict sense—it does not prescribe or allow a behavior—but rather a stipulative definition, or, in more general terms, a *meaning postulate* that constitutes the legal meaning of theft. Such sentences are often part of the legal code. They share with norms the property of being neither true nor false: stipulative definitions are neither empirical statements nor descriptive statements. In this sense we say that they are neither true nor false. However, they are held to be true by definition. The significance of (3) is that it decomposes the complex legal term 'theft' into more basic legal concepts. These concepts are again the subject of further meaning postulates, among which may be the following:

(4) A person in the sense of the law is a human being that has been born.
(5) A movable object is any physical object that is not a person or a piece of land.
(6) A movable object is in the possession of a person if that person is able to control the uses and the location of the object.
(7) The owner of an object is—within the limits of the law—entitled to do with it whatever he wants, namely keep it, use it, transfer possession or ownership of the object to another person, and destroy or abandon it.

Not all of definitions (4)-(7) may be found in the legal statutes, though they may be viewed as belonging to the normative system by virtue of having been accepted in legal theory and judicial reasoning. They constitute 'intermediate concepts': they link legal terms (person, movable object, possession etc.) to words describing natural facts (human being, born, piece of land, keep an object etc.).

Any proper representation of legal norms must include means of representing meaning postulates that define legal terms, decompose legal terms into more basic legal terms, or serve as intermediate concepts that link legal terms to terms that describe natural facts. But for deontic logic, with its standard possible worlds semantics, a comprehensive solution to the problem of representing meaning postulates is so far lacking (cf. Lindahl [78]).

Challenge 11. *How can meaning postulates and intermediate terms be modeled in semantics for deontic logic reasoning?*

The representation of intermediate concepts is of particular interest, since such concepts arguably reduce the number of implications required for the transition from natural facts to legal consequences and thus serve an economy of expression (cf. Lindahl and Odelstad [79] and their recent overview chapter [80]). Lindahl and Odelstad use the term 'ownership' as an example to argue as follows: let $F_1, ..., F_p$ be descriptions of some situations in which a person α acquires ownership of an object γ, e.g. by acquiring it from some other person β, finding it, building it from owned materials, etc., and let $C_1, ..., C_n$ be among the legal consequences of α's ownership of γ, e.g. freedom to use the object, rights to compensation when the object is damaged, obligations to maintain the object or pay taxes for it etc. To express that each fact F_i has the consequence C_j, $p \times n$ implications are required. The introduction of the term $Ownership(x, y)$ reduces the number of required implications to $p + n$: there are p implications that link the facts $F_1, ..., F_p$ to the legal term $Ownership(x, y)$, and n implications that link the legal term $Ownership(x, y)$ to each of the legal consequences $C_1, ..., C_n$. The argument obviously does not apply to all cases: one implication $(F_1 \vee ... \vee F_p) \to (C_1 \wedge ... \wedge C_n)$ may often be sufficient to represent the case that a variety of facts $F_1, ..., F_p$ has the same multitude of legal consequences $C_1, ..., C_n$. However, things may be different when norms that link a number of factual descriptions to the same legal consequences stem from different normative sources, may come into conflict with other norms, can be overridden by norms of higher priority, or be subject to individual exemption by norms that grant freedoms or licenses: in these cases, the norms must be represented individually. So it seems worthwhile to consider ways to incorporate intermediate concepts into a formal semantics for deontic logic.

In an input/output framework, a first step could be to employ a separate set T of theoretical terms, namely meaning postulates, alongside the set G of norms. Let T consists of intermediates of the form (a, x), where a is a factual sentence (e.g. that β is in possession of γ, and that α and β agreed that α should have γ, and that β hands γ to α), and x states that some legal term obtains (e.g. that α is now owner of γ). To derive outputs from the set of norms G, one may then use $A \cup out(T, A)$ as input, i.e. the factual descriptions together with the legal statements that obtain given the intermediates T and the facts A.

It may be of particular interest to see that such a set of intermediates may help resolve possible conflicts in the law. Let $(\top, \neg dog)$ be a statute that forbids dogs on the premises, but let there also be a higher order principle that no blind person may be required to give up his or her guide dog. Of course the conflict may be solved by modifying the statute (e.g. add a condition that the dog in question is not a guide dog), but then modifying a statute is usually not something a judge, faced with such a norm, is allowed to do: the judge's duty is solely to consider the statute, interpret it according to the known or supposed will of the norm-giver, and apply it to the given facts. The judge may then come to the conclusion that a fair and considerate norm-giver would not have meant the statute to apply to guide dogs, i.e. the term "dog" in the statute is a theoretical term whose extension is smaller than the natural term. So the statute must be re-interpreted as reading $(\top, \neg tdog)$ with the additional intermediate $(dog \wedge \neg guidedog, tdog) \in T$, and thus no conflict arises for the case of blind persons that want to keep their guide dog. While this seems to be a rather natural view of how judicial conflict resolution works (the example is taken from an actual court case), the exact process of creating and modifying theoretical terms in order to resolve conflicts must be left to further study.

12 Constitutive norms

Constitutive norms like counts-as conditionals are rules that create the possibility of or define an activity. For example, according to Searle [102], the activity of playing chess is constituted by action in accordance with these rules. Chess has no existence apart from these rules. The institutions of marriage, money, and promising are like the institutions of baseball and chess in that they are systems of such constitutive rules or conventions. They have been identified as the key mechanism to normative reasoning in dynamic and uncertain environments, for example to realize agent communication, electronic contracting, dynamics of organizations, see, e.g., Boella and van der Torre [14].

Challenge 12. *How to define counts-as conditionals and relate them to obligations and permissions?*

For Jones and Sergot [64], the counts-as relation expresses the fact that a state of affairs or an action of an agent "is a sufficient condition to guarantee that the institution creates some (usually normative) state of affairs". They formalize this introducing a conditional connective $\Rightarrow_s$ to express the "counts-as" connection in the context of an institution s. They characterize the logic of $\Rightarrow_s$ as a conditional logic, with axioms for agglomeration $((x \Rightarrow_s y) \& (x \Rightarrow_s z)) \supset (x \Rightarrow_s (y \wedge z))$, left disjunction $((x \Rightarrow_s z) \& (y \Rightarrow_s z)) \supset ((x \vee y) \Rightarrow_s z)$ together with transitivity $((x \Rightarrow_s y) \& (y \Rightarrow_s z)) \supset (x \Rightarrow_s z)$. The flat fragment can be phrased as an input/output logic as follows [15].

Definition 12.1. *Let L be a propositional action logic with $\vdash$ the related notion of derivability and Cn the related consequence operation $Cn(x) = \{y \mid x \vdash y\}$. Let CA be a set of pairs of L, $\{(x_1, y_1), \ldots, (x_n, y_n)\}$, read as '$x_1$ counts as y_1', etc. Moreover, consider the following proof rules conjunction for the output (AND), disjunction of the input (OR), and transitivity (T) defined as follows:*

$$\frac{(x, y_1), (x, y_2)}{(x, y_1 \wedge y_2)} AND \qquad \frac{(x_1, y), (x_2, y)}{(x_1 \vee x_2, y)} OR \qquad \frac{(x, y_1), (y_1, y_2)}{(x, y_2)} T$$

For an institution s, the counts-as output operator out_{CA} is defined as the closure operator on the set CA using the rules above together with a tacit rule that allows replacement of logical equivalents in input and output. We write $(x, y) \in out_{CA}(CA, s)$. Moreover, for $X \subseteq L$, we write $y \in out_{CA}(CA, s, X)$ if there is a finite $X' \subseteq X$ such that $(\wedge X', y) \in out_{CA}(CA, s)$, indicating that the output y is derived by the output operator for the input X, given the counts-as conditionals CA of institution s. We also write $out_{CA}(CA, s, x)$ for $out_{CA}(CA, s, \{x\})$.

Example 12.2. *If for some institution s we have $CA = \{(a, x), (x, y)\}$, then we have $out_{CA}(CA, s, a) = \{x, y\}$.*

The recognition that statements like "X counts as Y in context c" may have different meanings in different situations lead Grossi et al. [45, 46] to propose a family of operators capturing four notions of counts-as conditionals. Starting from a simple modal logic of contexts, several logics are used to define the family of operators. All logics have been proven to be sound and strongly complete. By using a logic of acceptance, Lorini et al. [81, 82] investigate another aspect of constitutive norms, that is, the fact that agents of a society need to accept such norms in order for them to be in force.

Considering the legal practice, Governatori and Rotolo [40] propose a study of constitutive norms within the framework of defeasible logic. This allows them to capture de defeasibility of counts-as conditionals: even in presence of a constitutive norms like "X counts as Y in context c", the inference of Y from X can be blocked in presence of exceptions.

There is presently no consensus on the logic of counts-as conditionals, probably due to the fact that the concept is not studied in depth yet. For example, the adoption of the transitivity rule T for their logic is criticized by Artosi et al. [8]. Jones and Sergot say that "we have been unable to produce any counter-instances [of transitivity], and we are inclined to accept it". Neither of these authors considers replacing transitivity by cumulative transitivity (CT): $((x \Rightarrow_s y) \& (x \wedge y \Rightarrow_s z)) \supset (x \Rightarrow_s z)$, that characterizes operations out_3, out_4 of input/output logic. For a more comprehensive overview on constitutive norms, the reader is referred to the chapter by Grossi and Jones [44] in the handbook of deontic logic and normative systems.

The main issue in defining constitutive norms like counts-as conditionals is defining their relation to regulative norms like obligations and permissions. Boella and van der Torre [15] use the notion of a logical architecture combining several logics into a more complex logical system, also called logical input/output nets (or *lions*).

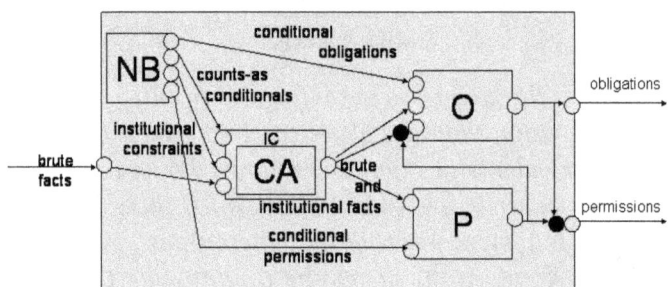

The notion of logical architecture naturally extends the input/output logic framework, since each input/output logic can be seen as the description of a 'black box'. In the above figure there are boxes for counts-as conditionals (CA), institutional constraints (IC), obligating norms (O) and explicit permissions (P). The norm base (NB) component contains sets of norms or rules, which are used in the other components to generate the component's output from its input. The figure shows that the counts-as conditionals are combined with the obligations and permissions using iteration, that is, the counts-as conditionals produce institutional facts, which are input for the norms. Roughly, if we write $out(CA, G, A)$ for the output of counts-as conditionals together with obligations, $out(G, A)$ for obligations as before, then $out(CA, G, A) = out(G, out_{CA}(CA, A))$.

There are many open issues concerning constitutive norms, since their logical analysis has not attracted much attention yet. How to distinguish among various kinds of constitutive norms? How are constitutive norms (x counts as y) distinguished from classifications (x is a y)? What is the relation with intermediate concepts?

13 Revision of a set of norms

In general, a code G of regulations is not static, but changes over time. For example, a legislative body may want to introduce new norms or to eliminate some existing ones. A different (but related) type of change is the one induced by the fusion of two (or more) codes—a topic addressed in the next section. A related but different issue not addressed here is that of how norms come about, how they propagate in the society, and how they change over time.

Little work exists on the logic of the revision of a set of norms. To the best of our knowledge, Alchourrón and Makinson [3, 4] were the first to study the changes of a legal code. The addition of a new norm n causes an enlargement of the code, consisting of the new norm plus all the regulations that can be derived from n. Alchourrón and Makinson distinguish two other types of change. When the new norm is incoherent with the existing ones, we have an *amendment* of the code: in order to coherently add the new regulation, we need to reject those norms that conflict with n. Finally, *derogation* is the elimination of a norm n together with whatever part of G implies n.

Alchourrón and Makinson [3] assume a "hierarchy of regulations". Alchourrón and Bulygin [2] also considered the *Normenordnung* and the consequences of gaps in this ordering. For example, in jurisprudence the existence of precedents is an established method to determine the ordering among norms.

However, although Alchourrón and Makinson aim at defining change operators for a set of norms of some legal system, the only condition they impose on G is that it is a non-empty and finite set of propositions. In other words, a norm x is taken to be simply a formula in propositional logic. Thus, they suggest that "the same concepts and techniques may be taken up in other areas, wherever problems akin to inconsistency and derogation arise" ([3], p. 147).

This explains how their work (together with Gärdenfors's analysis of counterfactuals) could ground that research area that is now known as *belief revision*. Belief revision is the formal study of how a set of propositions changes in view of new information that may be inconsistent with the existing beliefs. Expansion, revision and contraction are the three belief change operations that Alchourrón, Gärdenfors and Makinson identified in their approach (called AGM) and that have a clear correspondence with the changes on a system of norms we mentioned above.

Challenge 13. *How to revise a set of regulations or obligations?*

Recently, AGM theory has been reconsidered as a framework for norm change. However, beside syntactic approaches where norm change is performed directly on the set of norms (as in AGM), there are also proposals that appeared in the dynamic logic literature and that could be described as semantic approaches.

One example of this is the dynamic context logic proposed by Aucher et al. [9], where norm change is a form of model update. Point of depart is a dynamic variant of the logic of context used to study counts-as conditionals introduced by Grossi et al. [46]. Context expansion and context contraction operators are defined. Context expansion and context contraction represent the promulgation and the derogation of constitutive norms respectively. One of the advantages of this approach is that it can be used for the formal specification and verification of computational models of interactions based on norms.

A formal account clearly rooted in the legal practice is the one proposed by Governatori and Rotolo [41]. In particular, the removal of norms can be performed by annulment or by abrogation. The crucial difference between these two mechanisms is that annulment removes a norm from the code and all its effects (past and future) are cancelled. Abrogation, on the other hand, does not operate retroactively, and so it leaves the effects of an abrogated norm holding in the past.

It should then be clear that, in order to capture the difference between annulment and abrogation, the temporal dimension is pivotal. For this reason, Governatori and Rotolo's first attempt is to use theory revision in Defeasible Logic without temporal reasoning is unsuccessful as it cannot capture retroactivity. They the add a temporal dimension to Defeasible Logic to keep track of the changes in a normative system and to deal with retroactivity. Norms are represented along two temporal dimensions: the time of validity when the norm enters in the normative system and the time of effectiveness when the norm can produce legal effects. This leads to keep multiple versions of a normative system are needed. If Governatori and Rotolo [41] manage to capture the temporal dimension that plays a role in legal modifications, the resulting formalisation is rather complex.

To overcome such complexity without losing hold on the legal practice, Governatori et al. [42] explored three AGM-like contraction operators to remove rules, add exceptions and revise rule priorities.

Boella et al. [12] also use AGM theory, where propositional formulas are replaced by pairs of propositional formulas to represent rules, and the classical consequence operator Cn is replaced by an input/output logic. Within this framework, AGM contraction and revision of rules are studied. It is shown that results from belief base dynamics can be transferred to rule base dynamics. However, difficulties arise in the transfer of AGM theory change to rule change. In particular, it is shown that the six basic postulates of AGM contraction are consistent only for some input/output logics but not for others. Furthermore, it is shown how AGM rule revision can be defined in terms of AGM rule contraction using the Levi identity.

When we turn to a proper representation of norms, as in the input/output logic framework, the AGM principles thus prove to be too general to deal with the revision of a normative system. For example, one difference between revising a set of beliefs and revising a set of regulations is the following: when a new norm is added, coherence may be restored by modifying some of the existing norms, not necessarily retracting some of them. The following example clarifies this point:

Example. If we have $\{(\top, a), (a, b)\}$ and we have that c is an exception to the obligation to do b, then we need to retract (c, b). Two possible solutions are $\{(\neg c, a), (a, b)\}$ or $\{(\top, a), (a \wedge \neg c, b)\}$.

Stolpe [106] also combines input/output logic and AGM theory to propose an abstract model of norm change. Contraction is used to represent the derogation of a norm, that is, the elimination of a norm together with whatever part of the code that implies that norm. This is rendered as an AGM partial meet contraction with a selection function for a set of norms in input/output logic. Stolpe gives a complete AGM-style characterisation of the derogation operation. Revision, on the other hand, serves to study the amendment of a code, which happens when we wish to add a new norm which is incoherent with the existing ones. Amendment is defined as a norm revision obtained via the Levi identity.

Future research must investigate whether general patterns in the revision and contraction of norms exist and how to formalize them. Another open question is whether other logics can offer a general framework for modelling norm change. Finally, more case studies showing that formally defined operators serve for a conceptual analysis of normative change are needed.

14 Merging sets of norms

We now turn to another type of change, that is the aggregation of regulations. This problem has been only recently addressed in the literature and therefore the findings are still incomplete.

The first noticeable thing is the lack of general agreement about where the norms that are to be aggregated come from:

1. some papers focus on the merging of conflicting norms that belong to the same normative system [29];

2. other papers assume that the regulations to be fused belong to different systems [18]; and finally

3. some authors provide patterns of possible rules to be combined, and consider both cases 1. and 2. above [43].

The first situation seems to be more a matter of coherence of the whole system rather than a genuine problem of fusion of norms. However, such approaches have the merit to reveal the tight connections between fusion of norms, non-monotonic logics and defeasible deontic reasoning. The initial motivation for the study of belief revision was the ambition to model the revision of a set of regulations. In contrast to this, the generalization of belief revision to *belief merging* is primarily dictated by the goal to tackle the problem—arising in computer science—of combining information from different sources. The pieces of information are represented in a formal language and the aim is to merge them in an (ideally) unique knowledge base. See Konieczny and Grégoire [71] for a survey on logic-based approaches to information fusion.

Challenge 14. *Can the belief merging framework deal with the problem of merging sets of norms?*

If, following Alchourrón and Makinson, we assume that norms are unconditional, then we could expect to use standard merging operators to fuse sets of norms. Yet once we consider conditional norms, as in the input/output logic framework, problems arise again. Moreover, most of the fusion procedures proposed in the literature seem to be inadequate for the scope.

To see why this is the case, we need to explain the merging approach in a few words. Let us assume that we have a finite number of belief bases $K_1, K_2, \ldots, K_n$ to merge. IC is the belief base whose elements are the integrity constraints (i.e., any condition that we want the final outcome to satisfy). Given a multi-set $E = \{K_1, K_2, \ldots, K_n\}$ and IC, a merging operator $\mathcal{F}$ is a function that assigns a belief base to E and IC. Let $\mathcal{F}_{IC}(E)$ be the resulting collective base from the IC fusion on E.

Fusion operators come in two types: model-based and syntax-based. The idea of a model-based fusion operator is that models of $\mathcal{F}_{IC}(E)$ are models of IC, which are preferred according to some criterion depending on E. Usually the preference information takes the form of a total pre-order on the interpretations induced by a notion of distance $d(w, E)$ between an interpretation w and E.

Syntax-based merging operators are usually based on the selection of some consistent subsets of $\bigcup E$ [10, 70]. The bases K_i in E can be inconsistent and the result does not depend on the distribution of the well formed formulas over the members of the group. Konieczny [70] refers the term 'combination' to the syntax-based fusion operators to distinguish them from the model-based approaches.

Finally, the model-based aggregation operators for bases of equally reliable sources can be of two sorts. On the one hand, there are majoritarian operators that are based on a principle of distance-minimization [77]. On the other hand, there are egalitarian operators, which look at the distribution of the distances in E [69]. These two types of merging try to capture two intuitions that often guide the aggregation of individual preferences into a social one. One option is to let the majority decide the collective outcome, and the other possibility is to equally distribute the individual dissatisfaction.

Obviously, these intuitions may well serve in the aggregation of individual knowledge bases or individual preferences, but have nothing to say when we try to model the fusion of sets of norms. Hence, for this purpose, syntactic merging operators may be more appealing. Nevertheless, the selection of a coherent subset depends on additional information like an order of priority over the norms to be merged, or some other meta-principles.

The reader may wonder about the relationships between merging sets of norms and the revision of a normative system. In particular, one may speculate that Challenge 14 is not independent of Challenge 13, and that a positive answer to Challenge 14 implies an answer to 13. This is indeed an interesting question, but we believe that the answer to this question is not straightforward. Konieczny and Pino Pérez [72] have shown that there are close links between belief merging operators and belief revision ones. In particular, they show that an IC merging operator is an extension of an AGM revision operator. However, as we have seen, it is not clear whether IC merging operators could be properly used to study the merging of norms.

An alternative approach is to generalize existing belief change operators to merging rules. This is the approach followed by Booth et al. [18], where merging operators defined using a consolidation operation and possibilistic logic are applied to the aggregation of conditional norms in an input/output logic framework. However, at this preliminary stage, it is not clear whether such methodology is more fruitful for testing the flexibility of existing operators to tackle other problems than the ones they were created for, or if this approach can really shed some light on the new riddle at hand.

Grégoire [43] takes a different perspective. Here, real examples from the Belgian-French bilateral agreement preventing double taxation are considered. These are fitted into a taxonomy of the most common legal rules with exceptions, and the combination of each pair of norms is analyzed. Moreover, both the situations in which the regulations come from the same system and those in which they come from different ones are contemplated, and some general principles are derived. Finally, a merging operator for rules with abnormality propositions is proposed. A limitation of Grégoire's proposal is that only the aggregation of rules with the same consequence is taken into account and, in our opinion, this neglects other sorts of conflicts that may arise, as we see now.

Cholvy and Cuppens [29] also call for non-monotonic reasoning in the treatment of contradictions, and present a method for merging norms. The proposal assumes an order of priority among the norms to be merged and this order is used to resolve the incoherence. Even though this is quite a strong assumption, Cholvy and Cuppens's work takes into consideration a broader type of incoherence than Grégoire [43]. In their example, an organization that works with secret documents has two rules. R_1 is "It is obligatory that any document containing some secret information is kept in a safe, when nobody is using this document". R_2 is "If nobody has used a given document for five years, then it is obligatory to destroy this document by burning it". As they observe, in order to deduce that the two rules are conflicting, we need to introduce the constraint that keeping a document and destroying it are contradictory actions. That is, the notion of coherence between norms can involve information not given by any norms.

15 Games, norms and obligations

Deontic logic has been developed as a logic for practical reasoning, and normative systems are used to guide, control, or regulate desired system behaviour. This raises a number of questions. For example, how are deontic logic and the logic of normative systems related to alternative decision and agent interaction models such as BDI theory, decision theory, game theory, or social choice theory? Moreover, how can deontic logic be extended with cognitive concepts such as beliefs, desires, goals, intentions, and commitments? Though there have been a few efforts to base deontic logic with a logic of knowledge to define knowledge based obligation [92], or to extend deontic logic with BDI concepts [20], we believe that such extensions have not been fully explored yet. For example, Kolodny and MacFarlane [68] describe a decision problem involving miners, as well as several dialogues scenarios, which highlight the problems of normative reasoning with agents.

Maybe the most fundamental challenge has become apparent in this article. We discussed how deontic STIT logics are based on interactions of agents in games, and we discussed how norm based deontic logics have been developed on the basis of detachment. However, these two approaches have not been combined yet. So this is our final challenge in this article.

Challenge 15. *How can deontic logic be based on both norm and detachment, as well as decision and game theory?*

Norms and games have been related before. Lewis [76] introduced master-slave games and Bulygin [24] introduced Rex-minister-subject games in a discussion on the role of permissive norms in normative systems and deontic logic. Moreover, deontic logic has been used as an element in games to partially influence the behavior of individual agents [17]. Van der Torre [109] proposes games as the foundation of deontic logic. He illustrate the notion of a violation game using a metaphor from daily life. A person faces the parental problem of letting the son go to bed in time, or letting him make his homework. The mother is obliging her son to eat his vegetables. As illustrated in the first drawing of Figure 5, the son did what his mother asked him to do.

Figure 5: Conformance, violation, incentive, violation, negotiation (Drawings by Egberdien van der Torre), from [van der Torre, 2010].

However, in the second drawing his behavior has changed. The son does not like vegetables, and when the parents tell the boy to eat his vegetables, he just says "No!" At the third drawing, when the son's desire not to eat vegetables became stronger than his motivation to obey his parents, the parents adapted their strategy and introduced the use of incentives. They told their son, "if you empty your plate you will get a dessert", or sometimes, "if you don't finish your plate, you don't get a dessert." The boy has a desire to eat a dessert, and this desire is stronger than the desire not to eat vegetables, so he is eating his vegetables again. However, after some time we reach the fourth figure where the incentive no longer works. The boy starts to protest and to negotiate. In those cases, the parents sometimes decide that the son will get his dessert even without eating his vegetables, for example, because the child still has eaten at least some of them, or because it is his birthday, or simply because they are not in the mood to argue. As visualized in the fifth figure, this makes the boy very happy. It is precisely this aspect that characterizes a violation game. The violation does not follow necessarily from the norm, but is subject to exceptions and negotiation.

Figure 6 models this example by a standard extensive game tree. Let's look first at one moment in time. The child decides first whether to eat his vegetables or not. But in this decision, he takes the response of his parents into account. In other words, he has a model of how the parents will respond to his behavior. In the deontic logic we propose here, based on a violation game, it is obligatory to empty the plate when the boy expects that not eating his vegetables leads to violation, not when a violation logically follows. By the way, we identify the recognition of violation and the sanction in the example for illustrative purposes, in reality usually two distinct steps can be distinguished.

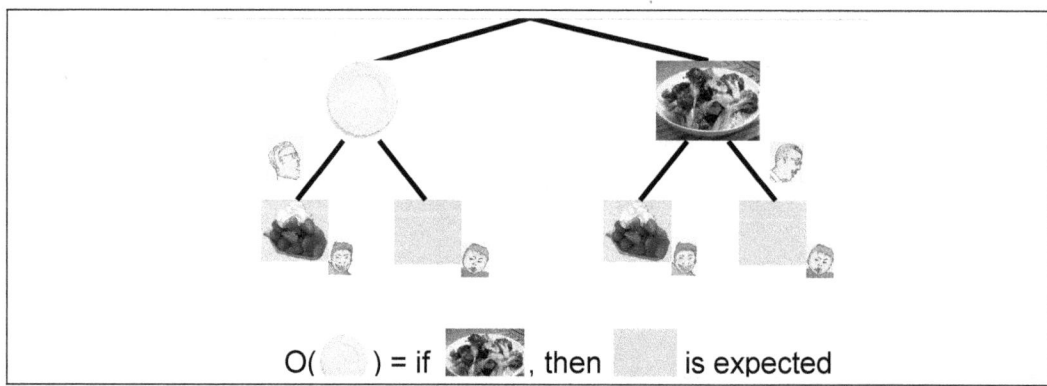

Figure 6: Expectation, from [van der Torre, 2010]

The general definition of obligation based on violation games extends this basic idea to behavior over periods of time. Let's consider the three phases in the example. Borrowing from terminology from classical game theory, we say that it is obligatory to eat the vegetables, when not eating them and the strategy that this leads to a violation, is an equilibrium. In the first phase in which the son eats his vegetables, the violation is only implicit since it does not occur. In the second phase not eating the vegetables is identified with the absence of the dessert. In the third phase, the boy may sometimes eat his vegetables, and sometimes not. As long as the norm is in force, he will still believe to be sanctioned most of the time when he does not eat his vegetables. When the sanction is not applied most of the time we have reached a fourth phase, in which we say that the norm is no longer in force.

Figure 7: Equilibrium, from [van der Torre, 2010]

Summarizing, norms are rules defining a violation game.

Definition 15.1 (Violation games [109]). *Violation games are social interactions among agents to determine whether violations have occurred, and which sanctions will be imposed for such violations. A normative system is a specification of violation games.*

Since norms do not have truth values, we cannot say that two normative systems are logically equivalent, or that a normative system implies a norm. Therefore it has been proposed to take equivalence of normative systems as the fundamental principle of deontic logic. Implication is then replaced by acceptance and redundancy, which are defined in terms of norm equivalence: a norm is accepted by a normative system if adding it to the normative system leads to an equivalent normative system, and a norm is redundant in a normative system if removing it from the normative system leads to an equivalent normative system. The fundamental notion of equivalence of normative systems can be defined in terms of violation games.

Definition 15.2 (Equivalence of normative systems [109]). *Two normative systems are equivalent if and only if they define the same set of violation games.*

Finally, we can now give a more precise definition of an autonomous system. Remember that auto means self, and nomos means norm.

Definition 15.3 (Autonomy [109]). *A system is autonomous if and only if it can play violation games.*

Violation games are the basis of normative reasoning and deontic logic, but more complex games must be considered too. Consider for example the following situation. If a child is in the water and there is one bystander, chances are that the bystander will jump into the water and save the child. However, if there are one hundred bystanders, chances are that no-one jumps in the water and the child will drown. How to reason about such bystander effects?

Van der Torre suggests that an extension of violation games, called norm creation games [17], may be used to analyze the situation. An agent reasons as follows. What is the explicit norm I would like to adopt for such situations? Clearly, if I would be in the water and I could not swim, or it is my child drowning in the water, then I would like prefer that someone would jump in the water. To be precise, I would accept a norm that in such cases, the norm for each individual would be to jump into the water. Consequently, one should act according to this norm, and everyone should jump into the water. Norm creation games can be used to give a more general definition of a normative system.

Definition 15.4 (Norm creation games [109]). *Norm creation games are social interactions among agents to determine which norms are in force, whether norm violations have occurred, and which sanctions will be imposed for such violations. A normative system is a specification of norm creation games.*

There are many details to be further discussed here. For example, if there is a way to discriminate among the people and it may be assumed that all people would follow this discrimination, then only some people have to jump into the water (the men, the good swimmers if they can be identified, the tall people, and so on). In general, and as common in legal reasoning, the more that is known about the situation, the more can be said about the protocol leading to the norm.

For the semantics of the new deontic logic founded on violation games, one needs a way to derive obligations from norms, as in the iterative detachment approach, or input/output logic. The extension now is to represent the agents and their games into the semantic structures, and derive the norms from that using game theoretic methods. As the norm creation game illustrates, also protocols for norm creation must be represented to model more complex games.

The language of the new deontic logic founded on violation games will be richer than most of the deontic logics studied thus far. There will be formal statements referring to the regulative, permissive and constitutive norms, as in the input/output logic framework, but there will also be an explicit representation of the games the agents are playing. Many choices are possible here, and the area of game theory will lead the way.

We need other approaches that represent norms and obligations at the same time, since deontic logic founded on violation games has to built on it. We also also have to study time, actions, mental modalities, permissions and constitutive norms, since they all play a role in violation games. We also need a precise understanding of Anderson's idea of violation conditions which do not necessarily lead to sanctions, but to the more abstract notion of "a bad state," i.e. a state in which something bad has happened. Whereas many of these deontic problems have been studied in isolation in the deontic logic literature, we believe that violation games will work as a metaphor to bring these problems together, and study their interdependencies.

16 Summary

The aim of this article is to introduce readers to the area of deontic logic and its challenges. The interested reader is advised to download the handbook of deontic logic and normative systems, and should not take our article only as its guidance. In particular, in this article we have not gone into the formal aspects of deontic logic. Deontic logicians have developed monadic modal logics, non-monotonic ones, rule based systems, and much more. The formalisms developed in deontic logic have also been adopted by a wider logic community, in particular the preference based deontic logics have been adopted in many areas [83].

As far as open problems are concerned, in the context of the handbook this concerns mainly the problems of *multiagent* deontic logic and problems related to normative systems. We have addressed the following challenges.

How to reconstruct the history of traditional deontic logic as a challenge to deal with contrary to duty reasoning, violations and preference (Challenge 1)?

What are the challenges in game theoretic approach to normative reasoning (Section 2), which is based on non-deterministic actions (Challenge 2), moral luck (Challenge 3) and procrastination (Challenge 4)?

How to reconstruct the history of modern deontic logic as a challenge to deal with Jørgensen's dilemma and detachment (Challenge 5), and more generally to bridge the tradition of normative system with the tradition of modal deontic logic?

What is the challenge in multi agent detachment of obligations from norms? For example, when detaching obligations from norms, when do agents assume that other agents comply with their norms (Challenge 6)? In game theory, agents assume that other agents are rational in the sense of acting in their best interest. Analogously, multiagent deontic logic raises the question when agents assume that other agents comply with their norms. For answering the question, we assume that every norm is directed towards a single agent, and that the normative system does not change.

How do norm based semantics handle the traditional challenges in deontic logic? These problems are when a set of norms may be termed 'coherent' (Challenge 7), how to deal with normative conflicts (Challenge 8), how to interpret dyadic deontic operators that formalize 'it ought to be that x on conditions α' as $O(x/\alpha)$ (Challenge 9), how various concepts of permission can be accommodated (Challenge 10), how meaning postulates and counts-as conditionals can be taken into account (Challenge 11 and 12), and how sets of norms may be revised and merged (Challenge 13 and 14).

Finally, how can the two approaches of game based deontic logic and norm based deontic logic be combined? (Challenge 15)

References

[1] Thomas Ågotnes, Wiebe van der Hoek, and Michael Wooldridge. Robust normative systems and a logic of norm compliance. *Logic Journal of the IGPL*, 18(1):4–30, 2010.

[2] C. E. Alchourrón and E. Bulygin. The expressive conception of norms. In R. Hilpinen, editor, *New Studies in Deontic Logic*, pages pp 95–124. Reidel, Dordrecht, 1981.

[3] C. E. Alchourrón and D. Makinson. Hierarchies of regulations and their logic. In R. Hilpinen, editor, *New Studies in Deontic Logic*, pages 125–148. Reidel, Dordrecht, 1981.

[4] C. E. Alchourrón and D. Makinson. On the logic of theory change: Contraction functions and their associated revision functions. *Theoria*, 48:14–37, 1982.

[5] A. R. Anderson. On the logic of commitment. *Philosophical Studies*, 19:23–27, 1959.

[6] G. Andrighetto, G. Governatori, P. Noriega, and L. van der Torre, editors. *Normative Multi-Agent Systems*, volume 4 of *Dagstuhl Follow-Ups*. Schloss Dagstuhl–Leibniz-Zentrum fuer Informatik, Dagstuhl, Germany, 2013.

[7] L. Åqvist. Good Samaritans, contrary-to-duty imperatives and epistemic obligations. *Noûs*, 1:361–379, 1967.

[8] A. Artosi, A. Rotolo, and S. Vida. On the logical nature of count-as conditionals. In *The Law of Electronic Agents. Proceedings of the LEA04 Workshop*, pages 9–33, Bologna, 2004.

[9] G. Aucher, D. Grossi, A. Herzig, and E. Lorini. Dynamic context logic. In X. He, J. Horty, and E. Pacuit, editors, *Logic, Rationality, and Interaction: Second International Workshop, LORI 2009, Chongqing, China, October 8-11, 2009. Proceedings*, pages 15–26, Berlin, Heidelberg, 2009. Springer Berlin Heidelberg.

[10] C. Baral, S. Kraus, J. Minker, and V. S. Subrahmanian. Combining knowledge bases consisting of first-order theories. *Computational Intelligence*, 8:45–71, 1992.

[11] G. Boella, J. M. Broersen, and L. van der Torre. Reasoning about constitutive norms, counts-as conditionals, institutions, deadlines and violations. In The Duy Bui, Tuong Vinh Ho, and Quang-Thuy Ha, editors, *Intelligent Agents and Multi-Agent*

Systems, 11th Pacific Rim International Conference on Multi-Agents, PRIMA 2008, Hanoi, Vietnam, December 15-16, 2008. Proceedings, volume 5357 of *Lecture Notes in Computer Science*, pages 86–97. Springer, 2008.

[12] G. Boella, G. Pigozzi, and L. van der Torre. AGM contraction and revision of rules. *Journal of Logic, Language and Information*, 25(3-4):273–297, 2016.

[13] G. Boella and L. van der Torre. Permissions and obligations in hierarchical normative systems. In *Proceedings of the 9th International Conference on Artificial Intelligence and Law, ICAIL 2003, Edinburgh, Scotland, UK, June 24-28, 2003*, pages 109–118, Edinburgh, 2003.

[14] G. Boella and L. van der Torre. Constitutive norms in the design of normative multiagent systems. In *Computational Logic in Multi-Agent Systems, 6th International Workshop, CLIMA VI*, LNCS 3900, pages 303–319. Springer, 2006.

[15] G. Boella and L. van der Torre. A logical architecture of a normative system. In *Deontic Logic and Artificial Normative Systems, 8th International Workshop on Deontic Logic in Computer Science (DEON'06)*, volume 4048 of *LNCS*, pages 24–35, Berlin, 2006. Springer.

[16] G. Boella, L. van der Torre, and H. Verhagen. Introduction to normative multiagent systems. *Computation and Mathematical Organizational Theory, special issue on normative multiagent systems*, 12(2-3):71–79, 2006.

[17] Guido Boella and Leendert van der Torre. A game-theoretic approach to normative multi-agent systems. In *Normative Multi-agent Systems, 18.03. - 23.03.2007*, Dagstuhl Seminar Proceedings, 2007.

[18] R. Booth, S. Kaci, and L. van der Torre. Merging rules: Preliminary version. In *Proceedings of the Eleventh International Workshop on Non-Monotonic Reasoning (NMR'06)*, 2006.

[19] C. Boutilier. Toward a logic for qualitative decision theory. In *Proceedings of the 4th International Conference on Principles of Knowledge Representation and Reasoning (KR'94). Bonn, Germany, May 24-27, 1994.*, pages 75–86, Bonn, 1994.

[20] J. Broersen, M. Dastani, and L. van der Torre. BDIOCTL: Obligations and the specification of agent behavior. In Georg Gottlob and Toby Walsh, editors, *IJCAI-03, Proceedings of the Eighteenth International Joint Conference on Artificial Intelligence, Acapulco, Mexico, August 9-15, 2003*, pages 1389–1390. Morgan Kaufmann, 2003.

[21] J. Broersen and L. van der Torre. Reasoning about norms, obligations, time and agents. In A. K. Ghose, G. Governatori, and R. Sadananda, editors, *Agent Computing and Multi-Agent Systems, 10th Pacific Rim International Conference on Multi-Agents, PRIMA 2007, Bangkok, Thailand, November 21-23, 2007. Revised Papers*, volume 5044 of *Lecture Notes in Computer Science*, pages 171–182. Springer, 2007.

[22] J. M. Broersen, F. Dignum, V. Dignum, and J-J Ch. Meyer. Designing a deontic logic of deadlines. In A. Lomuscio and D. Nute, editors, *Deontic Logic in Computer Science, 7th International Workshop on Deontic Logic in Computer Science, DEON 2004, Madeira, Portugal, May 26-28, 2004. Proceedings*, volume 3065 of *Lecture Notes in Computer Science*, pages 43–56. Springer, 2004.

[23] J. M. Broersen and L. van der Torre. What an agent ought to do. *Artif. Intell. Law*, 11(1):45–61, 2003.

[24] E. Bulygin. Permissive norms and normative systems. In A. Martino and F. Socci Natali, editors, *Automated Analysis of Legal Texts*, pages 211–218. Publishing Company, Amsterdam, 1986.

[25] J. Carmo and A.J.I. Jones. Deontic logic and contrary-to-duties. In D.Gabbay and F.Guenthner, editors, *Handbook of Philosophical Logic*, volume 8, pages 265–343. Kluwer, 2002.

[26] H-N Castañeda. The paradoxes of deontic logic: The simplest solution to all of them in one fell swoop. In R. Hilpinen, editor, *New Studies in Deontic Logic: Norms, Actions, and the Foundations of Ethics*, pages 37–85. Springer Netherlands, Dordrecht, 1981.

[27] H-N Castañeda. The logical structure of legal systems: A new perspective. In A. A. Martino, editor, *Deontic Logic, Computational Linguistics and Legal Information Systems, volume II*, page 21?37. North Holland Publishing, Dordrecht, 1982.

[28] R.M. Chisholm. Contrary-to-duty imperatives and deontic logic. *Analysis*, 24:33Ü36, 1963.

[29] L. Cholvy and F. Cuppens. Reasoning about norms provided by conflicting regulations. In P. McNamara and H. Prakken, editors, *Norms, Logics and Information Systems*, pages 247–264. IOS, Amsterdam, 1999.

[30] L. Cholvy and C. Garion. An attempt to adapt a logic of conditional preferences for reasoning with contrary-to-duties. *Fundam. Inform.*, 48(2-3):183–204, 2001.

[31] C. Condoravdi and S. Lauer. Anankastic conditionals are just conditionals. *Semantics and Pragmatics*, 9(8):1–69, November 2016.

[32] S. Danielsson. *Preference and Obligation: Studies in the Logic of Ethics*. Filosofiska fÅűreningen, Uppsala, 1968.

[33] J. Forrester. Gentle murder, or the adverbial Samaritan. *Journal of Philosophy*, 81:193–196, 1984.

[34] D. Gabbay, J. Horty, X. Parent, R. van der Meyden, and L. van der Torre, editors. *Handbook of Deontic Logic and Normative Systems*. College Publications, London, UK, 2013.

[35] L. Goble. Multiplex semantics for deontic logic. *Nordic Journal of Philosophical Logic*, 5:113–134, 2000.

[36] L. Goble. A logic for deontic dilemmas. *Journal of Applied Logic*, 3:461–483, 2005.

[37] L. Goble. Normative conflicts and the logic of 'ought'. *Noûs*, 43:450–489, 2009.

[38] L. Goble. Prima facie norms, normative conflicts, and dilemmas. In D. Gabbay, J. Horty, X. Parent, R. van der Meyden, and L. van der Torre, editors, *Handbook of Deontic Logic*, pages 249–352. College Publications, 2013.

[39] G. Governatori. Thou shalt is not you will. In *Proceedings of the 15th International Conference on Artificial Intelligence and Law, ICAIL 2015, San Diego, CA, USA, June 8-12, 2015*, pages 63–68, 2015.

[40] G. Governatori and A. Rotolo. A computational framework for institutional agency.

Artificial Intelligence and Law, 16(1):25–52, 2008.

[41] G. Governatori and A. Rotolo. Changing legal systems: legal abrogations and annulments in defeasible logic. *Logic Journal of IGPL*, 18(1):157–194, 2010.

[42] G. Governatori, A. Rotolo, F. Olivieri, and S. Scannapieco. Legal contractions: A logical analysis. In E. Francesconi and B. Verheij, editors, *Proceedings of the 9th International Conference on Artificial Intelligence and Law, ICAIL 2003, Edinburgh, Scotland, UK, June 24-28, 2003*, pages 63–72. ACM, 2013.

[43] E. Grégoire. Fusing legal knowledge. In *Proceedings of the 2004 IEEE Int. Conf. on Information Reuse and Integration (IEEE-IRI'2004)*, pages 522–529, 2004.

[44] D. Grossi and A. Jones. Constitutive norms and counts-as conditionals. In D. Gabbay, J. Horty, X. Parent, R. van der Meyden, and L. van der Torre, editors, *Handbook of Deontic Logic and Normative Systems*, pages 407–441. College Publications, 2013.

[45] D. Grossi, J-J Ch. Meyer, and F. Dignum. Classificatory aspects of counts-as: An analysis in modal logic. *Journal of Logic and Computation*, 16(5):613–643, 2006.

[46] D. Grossi, J-J Ch. Meyer, and F. Dignum. The many faces of counts-as: A formal analysis of constitutive-rules. *Journal of Applied Logic*, 6(2):192–217, 2008.

[47] J. Hansen. Sets, sentences, and some logics about imperatives. *Fundamenta Informaticae*, 48:205–226, 2001.

[48] J. Hansen. Problems and results for logics about imperatives. *Journal of Applied Logic*, 2:39–61, 2004.

[49] J. Hansen. Conflicting imperatives and dyadic deontic logic. *Journal of Applied Logic*, 3:484–511, 2005.

[50] J. Hansen. Deontic logics for prioritized imperatives. *Artificial Intelligence and Law*, 3(3-4):484–511, 2005.

[51] J. Hansen. Prioritized conditional imperatives: problems and a new proposal. *Autonomous Agents and Multi-Agent Systems*, 17(1):11–35, 2008.

[52] J. Hansen, G. Pigozzi, and L. van der Torre. Ten philosophical problems in deontic logic. In *Normative Multi-agent Systems, 18.03. - 23.03.2007*, Dagstuhl Seminar Proceedings, 2007.

[53] B. Hansson. An analysis of some deontic logics. *Noûs*, 3:373–398, 1969. Reprinted in [56] 121–147.

[54] S. O. Hansson. The varieties of permission. In D. Gabbay, J. Horty, X. Parent, R. van der Meyden, and L. van der Torre, editors, *Handbook of Deontic Logic and Normative Systems*, pages 195–240. College Publications, 2013.

[55] R. M. Hare. *Moral Thinking*. Clarendon Press, Oxford, 1981.

[56] R. Hilpinen, editor. *Deontic Logic: Introductory and Systematic Readings*. Reidel, Dordrecht, 1971.

[57] R. Hilpinen and P. McNamara. Deontic logic: A historical survey and introduction. In D. Gabbay, J. Horty, X. Parent, R. van der Meyden, and L. van der Torre, editors, *Handbook of Deontic Logic and Normative Systems*, pages 3–136. College Publications, 2013.

[58] J. Horty. *Agency and deontic logic.* Oxford University Press, 2001.

[59] J. F. Horty. Moral dilemmas and nonmonotonic logic. *Journal of Philosophical Logic,* 23:35–65, 1994.

[60] J. F. Horty. Nonmonotonic foundations for deontic logic. In D. Nute, editor, *Defeasible Deontic Logic,* pages 17–44. Kluwer, Dordrecht, 1997.

[61] J. F. Horty. Reasoning with moral conflicts. *Noûs,* 37:557–605, 2003.

[62] J. F. Horty. *Reasons and Defaults.* Oxford University Press, 2012.

[63] F. Jackson and R. Pargetter. Oughts, options, and actualism. *Philosophical Review,* 99:233–255, 1986.

[64] A. Jones and M. Sergot. A formal characterisation of institutionalised power. *Journal of IGPL,* 3:427–443, 1996.

[65] J. Jørgensen. Imperatives and logic. *Erkenntnis,* 7:288–296, 1938.

[66] H. Kamp. Free choice permission. *Proceedings of the Aristotelian Society,* 74:57–74, 1973.

[67] S. Kanger. New foundations for ethical theory: Part 1. duplic., 42 p., 1957. Reprinted in [56] 36–58.

[68] N. Kolodny and J. MacFarlane. Ifs and oughts. *Journal of Philosophy,* 107(3):115–143, 2010.

[69] S. Konieczny. *Sur la Logique du Changement: Révision et Fusion de Bases de Connaissance.* PhD thesis, University of Lille, France, 1999.

[70] S. Konieczny. On the difference between merging knowledge bases and combining them. In *KR 2000, Principles of Knowledge Representation and Reasoning Proceedings of the Seventh International Conference, Breckenridge, Colorado, USA, April 11-15, 2000.,* volume 8, pages 135–144. Morgan Kaufmann, 2000.

[71] S. Konieczny and E. Grégoire. Logic-based approaches to information fusion. *Information Fusion,* 7:4–18, 2006.

[72] S. Konieczny and Ramón P. Pérez. Logic based merging. *Journal of Philosophical Logic,* 40(2):239–270, 2011.

[73] E. J. Lemmon. Moral dilemmas. *The Philosophical Review,* 70:139–158, 1962.

[74] D. Lewis. *Counterfactuals.* Basil Blackwell, Oxford, 1973.

[75] D. Lewis. Semantic analyses for dyadic deontic logic. In S. Stenlund, editor, *Logical Theory and Semantic Analysis,* pages 1 – 14. Reidel, Dordrecht, 1974.

[76] D. Lewis. A problem with permission. In E. Saarinen, R. Hilpinen, I. Niiniluoto, and M. P. Hintikka, editors, *Essays in Honour of Jaako Hintikka: On the Occasion of His Fiftieth Birthday on January 12, 1979,* pages 163–175. Reidel, Dordrecht, 1979.

[77] J. Lin and A. Mendelzon. Merging databases under constraints. *International Journal of Cooperative Information Systems,* 7:55–76, 1996.

[78] L. Lindahl. Norms, meaning postulates, and legal predicates. In E. Garzón Valdés, editor, *Normative Systems in Legal and Moral Theory. Festschrift for Carlos E. Alchourrón and Eugenio Bulygin,* pages 293–307. Duncker & Humblot, Berlin, 1997.

[79] L. Lindahl and J. Odelstad. Intermediate concepts in normative systems. In L. Goble and J-J Ch. Meyer, editors, *Deontic Logic and Artificial Normative Systems: 8th International Workshop on Deontic Logic in Computer Science, DEON 2006, Utrecht, The Netherlands, July 12-14, 2006. Proceedings*, pages 187–200. Springer Berlin Heidelberg, Berlin, Heidelberg, 2006.

[80] L. Lindahl and J. Odelstad. The theory of joining-systems. In D. Gabbay, J. Horty, X. Parent, R. van der Meyden, and L. van der Torre, editors, *Handbook of Deontic Logic and Normative Systems*, pages 545–634. College Publications, 2013.

[81] E. Lorini and D. Longin. A logical account of institutions: From acceptances to norms via legislators. In *Proceedings of the Eleventh International Conference on Principles of Knowledge Representation and Reasoning*, KR'08, pages 38–48. AAAI Press, 2008.

[82] E. Lorini, D. Longin, B. Gaudou, and A. Herzig. The logic of acceptance: Grounding institutions on agents' attitudes. *Journal of Logic and Computation*, 19(6):901–940, 2009.

[83] D. Makinson. Five faces of minimality. *Studia Logica*, 52:339–379, 1993.

[84] D. Makinson. On a fundamental problem of deontic logic. In P. McNamara and H. Prakken, editors, *Norms, Logics and Information Systems. New Studies on Deontic Logic and Computer Science*, pages 29–54. IOS Press, 1999.

[85] D. Makinson and L. van der Torre. Input-output logics. *Journal of Philosophical Logic*, 29(4):383–408, 2000.

[86] D. Makinson and L. van der Torre. Constraints for input-output logics. *Journal of Philosophical Logic*, 30(2):155–185, 2001.

[87] D. Makinson and L. van der Torre. Permissions from an input-output perspective. *Journal of Philosophical Logic*, 32(4):391–416, 2003.

[88] D. Makinson and L. van der Torre. What is input/output logic? In B. Löwe, W. Malzkom, and T. Räsch, editors, *Foundations of the Formal Sciences II : Applications of Mathematical Logic in Philosophy and Linguistics (Papers of a conference held in Bonn, November 10-13, 2000)*, Trends in Logic, vol. 17, pages 163–174, Dordrecht, 2003. Kluwer. Reprinted in this volume.

[89] R. B. Marcus. Moral dilemmas and consistency. *Journal of Philosophy*, 77:121?136, 1980.

[90] I. Niiniluoto. Hypothetical imperatives and conditional obligation. *Synthese*, 66:111–133, 1986.

[91] Loes Olde Loohuis. Obligations in a responsible world. In H. Xiangdong, J. F. Horty, and E. Pacuit, editors, *LORI*, volume 5834 of *Lecture Notes in Computer Science*, pages 251–262. Springer, 2009.

[92] E. Pacuit, R. Parikh, and E. Cogan. The logic of knowledge based obligation. *Synthese*, 149(2):311–341, 2006.

[93] X. Parent. On the strong completeness of åqvist's dyadic deontic logic G. In R. van der Meyden and L. van der Torre, editors, *Deontic Logic in Computer Science, 9th International Conference, DEON 2008, Luxembourg, Luxembourg, July 15-18, 2008. Pro-

ceedings, volume 5076 of *Lecture Notes in Computer Science*, pages 189–202. Springer, 2008.

[94] X. Parent and L. van der Torre. Aggregative deontic detachment for normative reasoning. In C. Baral, G. De Giacomo, and T. Eiter, editors, *Principles of Knowledge Representation and Reasoning: Proceedings of the Fourteenth International Conference, KR 2014, Vienna, Austria, July 20-24, 2014*. AAAI Press, 2014.

[95] X. Parent and L. van der Torre. "sing and dance!" - input/output logics without weakening. In *Deontic Logic and Normative Systems - 12th International Conference, DEON 2014, Ghent, Belgium, July 12-15, 2014. Proceedings*, pages 149–165, 2014.

[96] X. Parent and L van der Torre. The pragmatic oddity in norm-based deontic logics. In *Proceedings of The 16th International Conference on Artificial Intelligence and Law*, 2017.

[97] J. Pearl. From conditional oughts to qualitative decision theory. In D. Heckerman and E. H. Mamdani, editors, *UAI '93: Proceedings of the Ninth Annual Conference on Uncertainty in Artificial Intelligence, The Catholic University of America, Providence, Washington, DC, USA, July 9-11, 1993*, pages 12–22. Morgan Kaufmann, 1993.

[98] H. Prakken and M. Sergot. Contrary-to-duty obligations. *Studia Logica*, 57:91–115, 1996.

[99] A. Ross. Imperatives and logic. *Theoria*, 7:53–71, 1941. Reprinted in *Philosophy of Science* **11**:30–46, 1944.

[100] W. D. Ross. *The Right and the Good*. Clarendon Press, Oxford, 1930.

[101] J.-P. Sartre. *L'Existentialisme est un Humanisme*. Nagel, Paris, 1946.

[102] J.R. Searle. *The Construction of Social Reality*. The Free Press, New York, 1995.

[103] T. J. Smiley. The logical basis of ethics. *Acta Philosophica Fennica*, 16:237–246, 1963.

[104] W. Spohn. An analysis of Hansson's dyadic deontic logic. *Journal of Philosophical Logic*, 4:237–252, 1975.

[105] E. Stenius. The principles of a logic of normative systems. *Acta Philosophica Fennica*, 16:247–260, 1963.

[106] A. Stolpe. Norm-system revision: Theory and application. *Artificial Intelligence and Law*, 18:247–283, 2010.

[107] X. Sun and L. van der Torre. Combining constitutive and regulative norms in input/output logic. In *Deontic Logic and Normative Systems - 12th International Conference, DEON 2014, Ghent, Belgium, July 12-15, 2014. Proceedings*, pages 241–257, 2014.

[108] J. van Benthem, D. Grossi, and F. Liu. Priority structures in deontic logic. *Theoria*, 80(2):116–152, 2014.

[109] L. van der Torre. Violation games: a new foundation for deontic logic. *Journal of Applied Non-Classical Logics*, 20(4):457–477, 2010.

[110] L. van der Torre and Y. Tan. The temporal analysis of chisholm's paradox. In J. Mostow and C. Rich, editors, *Proceedings of the Fifteenth National Conference on Artificial Intelligence and Tenth Innovative Applications of Artificial Intelligence*

Conference, AAAI 98, IAAI 98, July 26-30, 1998, Madison, Wisconsin, USA., pages 650–655. AAAI Press / The MIT Press, 1998.

[111] L. van der Torre and Y. Tan. An update semantics for prima facie obligations. In *Proceedings of The 17th European Conference on Artificial Intelligence*, pages 38–42, 1998.

[112] L. van der Torre and Y. Tan. Rights, duties and commitments between agents. In T. Dean, editor, *Proceedings of the Sixteenth International Joint Conference on Artificial Intelligence, IJCAI 99, Stockholm, Sweden, July 31 - August 6, 1999. 2 Volumes, 1450 pages*, pages 1239–1246. Morgan Kaufmann, 1999.

[113] L. van der Torre and Y. Tan. An update semantics for defeasible obligations. In K. B. Laskey and H. Prade, editors, *UAI '99: Proceedings of the Fifteenth Conference on Uncertainty in Artificial Intelligence, Stockholm, Sweden, July 30 - August 1, 1999*, pages 631–638. Morgan Kaufmann, 1999.

[114] B. van Fraassen. Values and the heart's command. *Journal of Philosophy*, 70:5–19, 1973.

[115] B. C. van Fraassen. The logic of conditional obligation. *Journal of Philosophical Logic*, 1:417–438, 1972.

[116] G. H. von Wright. Deontic logic. *Mind*, 60:1–15, 1951.

[117] G. H. von Wright. *An Essay in Modal Logic*. North-Holland, Amsterdam, 1951.

[118] G. H. von Wright. A note on deontic logic and derived obligation. *Mind*, 65:507–509, 1956.

[119] G. H. von Wright. A new system of deontic logic. *Danish Yearbook of Philosophy*, 1:173–182, 1961. Reprinted in [56] 105–115.

[120] G. H. von Wright. A correction to a new system of deontic logic. *Danish Yearbook of Philosophy*, 2:103–107, 1962. Reprinted in [56] 115–119.

[121] G. H. von Wright. *Norm and Action*. Routledge & Kegan Paul, London, 1963.

[122] G. H. von Wright. *An Essay in Deontic Logic and the General Theory of Action*. North Holland, Amsterdam, 1968.

[123] G. H. von Wright. Norms, truth and logic. In G. H. von Wright, editor, *Practical Reason: Philosophical Papers vol. I*, pages 130–209. Blackwell, Oxford, 1983.

[124] G.H von Wright. *Logical Studies*. Routledge and Kegan, London, 1957.

[125] Z. Ziemba. Deontic syllogistics. *Studia Logica*, 28:139–159, 1971.

Received 30 October 2016

Detachment in Normative Systems: Examples, Inference Patterns, Properties

Xavier Parent, Leendert van der Torre
University of Luxembourg
{xavier.parent,leon.vandertorre}@uni.lu

Abstract

There is a variety of ways to reason with normative systems. This partly reflects a variety of semantics developed for deontic logic, such as traditional semantics based on possible worlds, or alternative semantics based on algebraic methods, explicit norms or techniques from non-monotonic logic. This diversity raises the question how these reasoning methods are related, and which reasoning method should be chosen for a particular application. In this paper we discuss the use of examples, inference patterns, and more abstract properties. First, benchmark examples can be used to compare ways to reason with normative systems. We give an overview of several benchmark examples of normative reasoning and deontic logic: van Fraassen's paradox, Forrester's paradox, Prakken and Sergot's cottage regulations, Jeffrey's disarmament example, Chisholm's paradox, Makinson's Möbius strip, and Horty's priority examples. Moreover, we distinguish various interpretations that can be given to these benchmark examples, such as consistent interpretations, dilemma interpretations, and violability interpretations. Second, inference patterns can be used to compare different ways to reason with normative systems. Instead of analysing the benchmark examples semantically, as it is usually done, in this paper we use inference patterns to analyse them at a higher level of abstraction. We discuss inference patterns reflecting typical logical properties such as strengthening of the antecedent or weakening of the consequent. Third, more abstract properties can be defined to compare different ways to reason with normative systems. To define these more abstract properties, we first present a formal framework around the notion of detachment. Some of the ten properties we introduce are derived from the

---

Thanks to an anonymous reviewer for valuable comments. This work is supported by the European Union's Horizon 2020 research and innovation programme under the Marie Curie grant agreement No: 690974 (Mining and Reasoning with Legal Texts, MIREL).

inference patterns, but others are more abstract: factual detachment, violation detection, substitution, replacements of equivalents, implication, para-consistency, conjunction, factual monotony, norm monotony, and norm induction. We consider these ten properties as desirable for a reasoning method for normative systems, and thus they can be used also as requirements for the further development of formal methods for normative systems and deontic logic.

Keywords: Deontic Logic, Normative Systems, Benchmarks, Inference Patterns, Framework, Properties

1 Introduction

The *Handbook of Deontic Logic and Normative Systems* [5] describes a debate between the traditional or standard semantics for deontic logic and alternative approaches. The traditional semantics is based on possible world models, whereas many alternative approaches refer to foundations in normative systems, algebraic methods, or non-monotonic logic. In particular, whereas Anderson [1] argued to refer explicitly to normative systems and also Åqvist [2] builds on it, various alternative approaches such as input/output logic [13, 14] represent norms explicitly in the semantics.

Proponents of alternative approaches typically refer to limitations in the traditional approach, although the traditional approach has been generalised or extended to handle many of these limitations [10]. The development of formal and conceptual bridges between traditional and alternative approaches is one of the main current challenges in the area of normative systems and deontic logic. The following three limitations are frequently discussed.

Dilemmas. Examples discussed in the literature are those of van Fraassen [30], Makinson [13]'s Möbius strip, Prakken and Sergot [20]'s cottage regulations, and Horty [9]'s priority examples.

Defeasibility. The traditional approach does not distinguish various kinds of defeasibility. Legal norms are often assumed to be defeasible, and there is an increasing interest in philosophy in defeasibility, such as the defeasibility of moral reasons [9, 16].

Identity. Many traditional deontic logics validate the formula $\bigcirc(\alpha|\alpha)$, read as "α is obligatory given α," "whose intuitive standing is open to question" [13]. This has been dismissed as a harmless borderline case by proponents of the traditional semantics,

but it hinders the representation of fulfilled obligations and violations, playing a central role in normative reasoning. Consider a logic validating identity: the formula $\bigcirc(\alpha|\neg\alpha)$, which represents explicitly that there is a violation, is not satisfiable; the obligation of α disappears, in context $\neg\alpha$. (See Section 2 in this paper.)

Different disciplines and applications have put forward different requirements for the development of formal methods for normative systems and deontic logic. For example, in linguistics compositionality is an important requirement, as deontic statements must be integrated into a larger theory of language. In legal informatics, constitutive and permissive norms play a central role, and legal norms may conflict. It is an open problem whether there can be a unique formal method which can be widely applied across disciplines, or even whether there is a single framework of formal methods which can be used. In this sense, there may be an important distinction between classical and normative reasoning, since there is a unique first order logic for classical logic reasoning about the real world using sets, relations and functions. The situation for normative reasoning may be closer to the situation for non-monotonic reasoning, where also a family of reasoning methods have been proposed, rather than a unique method.

In this paper we do not want to take a stance on these discussions, but we want to provide techniques and ideas to compare traditional and alternative approaches. We focus on inference patterns and proof-theory instead of semantical considerations. In particular, in this paper we are interested in the question:

> Which obligations can be detached from a set of rules or conditional norms in a context?

Our angle is different from the more traditional one in terms of inference rules.

There are many frameworks for reasoning about rules and norms, and there are many examples about detachment from normative systems, many of them problematic in some sense. However, there are few properties to compare and analyse ways to detach obligations from rules and norms, and they are scattered over the literature. We are not aware of a systematic overview of these properties. We address our research question by surveying examples, inference patterns and properties from the deontic logic literature.

Examples: Van Fraassen's paradox, Forrester's paradox, Prakken and Sergot's cottage regulations, Jeffrey's disarmament example, Chisholm's paradox, Makinson's Möbius strip, and Horty's priority example. They illustrate challenges for normative reasoning with deontic dilemmas, contrary-to-duty reasoning, defeasible obligations, reasoning by cases, deontic detachment, prioritised obligations, and combinations of these.

Inference Patterns: Conjunction, weakening of the consequent, forbidden conflict, factual detachment, strengthening of the antecedent, violation detection, compliance detection, reinstatement, deontic detachment, transitivity, and various variants of these patterns.

Framework: We develop a *framework* for deontic logics representing and resolving conflicts. By framework we mean that we do not develop a single logic, but many of them. This reflects that there is not a single logic of obligation and permission, but many of them, and which one is to be used depends on the application.

Properties: Factual detachment, violation detection, substitution, replacements of equivalents, implication, paraconsistency, conjunction, factual monotony, norm monotony, and norm induction.

The term "property" is more general than the term "inference pattern". An inference pattern describes a property of a certain form. The inference patterns listed above appear also in the list properties. For instance, factual monotony echoes strengthening of the antecedent. In some cases, we use the same name for both the inference pattern and the corresponding property.

A formal framework to compare formal methods should make as little assumptions as possible, so it is widely applicable. We only assume that the context is a set of facts $\{a, b, \ldots\}$ and that the conditional norms are of the type "if a is the case, then it ought to be the case that b" where a and b are sentences of a propositional language. This is more general than some rule-based languages based on logic programming, where a is restricted to a conjunction of literals and b is a single literal. However, it is less expressive than many other languages, that contain, for example, modal or first order sentences, constitutive and permissive norms, mixed norms such as "if a is permitted, then b is obligatory," nested operators, time, actions, knowledge, and so on. There are few benchmark examples discussed in the literature for such an extended language (see [6] for a noteworthy exception) and we are not aware of any properties specific for such extended languages. Extending our formal framework and properties to such extended languages is therefore left to further research.

Our framework is built upon the notion of detachment. In traditional approaches "if a, then it ought that b" is typically written as either $a \to \bigcirc b$ or as $\bigcirc(b|a)$, and in alternative approaches it is sometimes written as (a, b). To be able to compare the different reasoning methods, we will not distinguish between these ways to represent normative systems. The challenge for comparing the formal approaches is that traditional methods typically derive conditional obligations, whereas alternative methods typically do not, maybe because they assume norms do not have truth values and thus they cannot be derived from other norms. Instead, they derive only unconditional obligations. To compare these approaches, one may assume that the derivation of a conditional obligation "if a, then it ought that b" is short for "if the context is exactly $\{a\}$, then the obligation $\bigcirc b$ is detached." Alternatively, the detachment of an obligation for b in context a in alternative systems may be written as the derivation of a pair (a, b), as it is done in the proof theory of input/output logics [13, 14]. These issues are discussed in more detail in Section 3 of this paper.

A remark on notation and terminology. We use Greek letters $\alpha, \beta, \gamma, \ldots$ for propositional formulas, and roman letters $a, b, c, \ldots, p, q, \ldots$ for (distinct) propositional atoms. Throughout this paper the terms "rule" and "conditional norm" will be used interchangeably. The term "rule" is most often used in computer science (with reference to so-called rule-systems and expert systems), and the term "conditional norm" in philosophy and linguistics. Readers should feel free to use the term they prefer. The unconditional obligation for α will be written as $\bigcirc \alpha$, while the conditional obligation for α given β will be written as $O(\alpha|\beta)$, or as (β, α). We do not assume a specific semantics for these constructs.

We give two examples below.

Example 1.1 (Deontic explosion). *The deontic explosion requirement says that we should not derive all obligations from a dilemma. Now consider a dilemma with obligations for $\alpha \wedge \beta$ and $\neg \alpha \wedge \gamma$. It may be tempting to think that an obligation for $\beta \wedge \gamma$ should follow:*

$$\frac{\dfrac{\bigcirc(\alpha \wedge \beta)}{\bigcirc \beta} \quad \dfrac{\bigcirc(\neg \alpha \wedge \gamma)}{\bigcirc \gamma}}{\bigcirc(\beta \wedge \gamma)}$$

Assuming that we have replacements by logical equivalents, if we substitute a for α, $a \vee b$ for β, and $\neg a \vee b$ for γ, then we would derive from the obligations for a and $\neg a$ the obligation for c: deontic explosion. We should not derive the obligation for $\beta \wedge \gamma$, because $\alpha \wedge \beta$ and $\neg \alpha \wedge \gamma$ are classically inconsistent. As we show in Section 2.1, the obligation for $\beta \wedge \gamma$ should be derived only under suitable assumptions.

Example 1.2 (Aggregation). *Consider an iterative approach deriving from the two norms "obligatory c given $a \wedge b$" and "obligatory b given a" that in some sense we have in context a that c is obligatory. This derivation of the obligation for c is made by so-called deontic* *detachment, because it is derived from the fact a together with the obligation for b. However, if the input is a together with the negation of b, then (intuitively) c should not be derived. However, we can (still intuitively) make the following two derivations. First, we can derive "obligatory a and b given c," a norm which is accepted by the two norms (Parent and van der Torre [18, 19]).*

$$\frac{O(\alpha|\beta \wedge \gamma), O(\beta|\gamma)}{O(\alpha \wedge \beta|\gamma)} \qquad \frac{(\gamma, \beta), (\gamma \wedge \beta, \alpha)}{(\gamma, \beta \wedge \alpha)}$$

Second, we can also derive the ternary norm "given α, and assuming β, γ is obligatory." However, we would need to extend the language with such expressions as done by van der Torre [27] and Xin & van der Torre [24]. Different motivations for using a ternary operator can be given. For instance, one may want to reason about exceptions to norms. This is the approach taken by van der Torre [27], who works with expressions of the form "given α, γ is obligatory unless β."

This paper is organised as follows. In Section 2 we introduce benchmark examples of deontic logic, and discuss them using inference patterns. In Section 3, we introduce the formal framework and its properties. Our approach is general and conceptual, and we abstract away from any specific system from literature. The reader will find in the *Handbook of Deontic Logic and Normative Systems* sample systems which can serve to exemplify the general considerations offered in this paper.

The present paper does not cover the notion of permission nor does it cover the notion of counts-as conditional. These topics will be a subject for future research. The reader is referred to the chapter by S. O. Hansson and to the chapter by A. Jones and D. Grossi in the aforementioned handbook for an overview of the state-of-the-art and perspectives for future research regarding these notions.

2 Benchmark Examples and Inference Patterns

In this section we discuss benchmark examples of deontic logic. The analysis in this section is based on a number of inference patterns. We do not consider ways in which deontic statements can be given a semantics. These principles must be understood as expressing strict rules. For future reference, we list the inference patterns in Table 1, in the order they are discussed in this section.

pattern	name
$\bigcirc \alpha_1, \bigcirc \alpha_2 / \bigcirc(\alpha_1 \wedge \alpha_2)$	AND
$\bigcirc \alpha_1, \bigcirc \alpha_2, \Diamond(\alpha_1 \wedge \alpha_2) / \bigcirc(\alpha_1 \wedge \alpha_2)$	RAND
$\bigcirc \alpha_1 / \bigcirc(\alpha_1 \vee \alpha_2)$	W
$\bigcirc(\alpha_1|\beta), \bigcirc(\alpha_2|\beta), \Diamond(\alpha_1 \wedge \alpha_2) / \bigcirc(\alpha_1 \wedge \alpha_2|\beta)$	RANDC
$\bigcirc(\alpha_1|\beta) / \bigcirc(\alpha_1 \vee \alpha_2|\beta)$	WC
$\bigcirc(\alpha_1|\beta), \bigcirc(\alpha_2|\beta), \Diamond(\alpha_1 \wedge \alpha_2 \wedge \beta) / \bigcirc(\alpha_1 \wedge \alpha_2|\beta)$	RANDC2
$\bigcirc(\alpha_1 \wedge \alpha_2|\beta_1), \bigcirc(\neg \alpha_1 \wedge \alpha_3|\beta_1 \wedge \beta_2) / \bigcirc(\neg \beta_2|\beta_1)$	FC
$\bigcirc(\alpha|\beta), \beta / \bigcirc \alpha$	FD
$\bigcirc(\alpha|\beta_1) / \bigcirc(\alpha|\beta_1 \wedge \beta_2)$	SA
$\bigcirc(\alpha|\beta_1), \Diamond(\alpha \wedge \beta_1 \wedge \beta_2) / \bigcirc(\alpha|\beta_1 \wedge \beta_2)$	RSA
$\bigcirc(\alpha|\beta) / \bigcirc(\alpha|\beta \wedge \neg \alpha)$	VD
$\bigcirc(\alpha|\beta \wedge \neg \alpha) / \bigcirc(\alpha|\beta)$	VD$^-$
$\bigcirc(\alpha|\beta_1), C / \bigcirc(\alpha|\beta_1 \wedge \beta_2)$	RSA$_C$
$\bigcirc(\alpha|\beta) / \bigcirc(\alpha|\beta \wedge \alpha)$	CD
$\bigcirc(\alpha|\beta \wedge \alpha) / \bigcirc(\alpha|\beta)$	CD
$\bigcirc(\alpha_1|\beta_1), \bigcirc(\neg \alpha_1 \wedge \alpha_2|\beta_1 \wedge \beta_2) / \bigcirc(\alpha_1|\beta_1 \wedge \beta_2 \wedge \neg \alpha_2)$	RI
$\bigcirc(\alpha_1|\beta_1), \bigcirc(\neg \alpha_1 \wedge \alpha_2|\beta_1 \wedge \beta_2),$ $\bigcirc(\neg \alpha_2|\beta_1 \wedge \beta_2 \wedge \beta_3) / \bigcirc(\alpha_1|\beta_1 \wedge \beta_2 \wedge \beta_3)$	RIO
$\bigcirc(\alpha|\beta_1), \bigcirc(\alpha|\beta_2) / \bigcirc(\alpha|\beta_1 \vee \beta_2)$	ORA
$\bigcirc(\alpha|\beta), \bigcirc \beta / \bigcirc \alpha$	DD
$\bigcirc(\alpha|\beta), \bigcirc(\beta|\gamma) / \bigcirc(\alpha|\gamma)$	T
$\bigcirc(\alpha|\beta \wedge \gamma), \bigcirc(\beta|\gamma) / \bigcirc(\alpha|\gamma)$	CT
$\bigcirc(\alpha|\beta \wedge \gamma), \bigcirc(\beta|\gamma) / \bigcirc(\alpha \wedge \beta|\gamma)$	ACT

Table 1: Inference patterns

The letter C in RSA$_C$ stands for the condition: there is no premise $\bigcirc(\alpha' \mid \beta')$ such that $\beta_1 \wedge \beta_2$ logically implies β', β' logically implies β_1 and not vice versa, α and α' are contradictory and $\alpha \wedge \beta'$ is consistent. RSA$_C$ is not a rule in the usual proof-theoretic sense. For it has a statement that quantifies over all other premises as an auxiliary condition. Thus the rule is not on a par with the other rules, like for instance weakening of the output.

2.1 Van Fraassen's Paradox

We first discuss deontic explosion in van Fraassen's paradox, then the trade-off between on the one hand "ought implies can" and on the other hand the representation of violations in the violation detection problem, whether it is forbidden to put oneself into a dilemma, and finally the use of priorities to resolve conflicts.

2.1.1 Deontic Explosion: Conjunction versus Weakening

It is a well-known problem from paraconsistent logic that the removal of all inconsistent formulas from the language is insufficient to reason in the presence of a contradiction, because there may still be explosion in the sense that all formulas of the language are derived from a contradiction. The following derivation illustrates how we can derive q from p and $\neg p$ in propositional logic, where all formulas in the derivation are classically consistent.

$$\frac{\dfrac{\dfrac{p}{q \vee p} \quad \neg p}{q \wedge \neg p}}{q}$$

The rules of replacements of logical equivalents, $\vee$-introduction, $\wedge$-introduction, and $\wedge$-elimination are used in this derivation.

A similar phenomenon occurs in deontic logic, if we reason about deontic dilemmas or conflicts, that is situations where $\bigcirc p$ and $\bigcirc \neg p$ both hold. Van der Torre and Tan [29] call this deontic explosion problem "van Fraassen's paradox," because van Fraassen [30] gave the following (informal) analysis of dilemmas in deontic logic. He rejects the conjunction pattern AND:

$$\text{AND:} \frac{\bigcirc \alpha_1, \bigcirc \alpha_2}{\bigcirc(\alpha_1 \wedge \alpha_2)}$$

This is because AND warrants the move from $\bigcirc p \wedge \bigcirc \neg p$ to $\bigcirc(p \wedge \neg p)$, and such a conclusion is not consistent with the principle 'ought implies can', formalised as $\neg \bigcirc (p \wedge \neg p)$. However, he does not want to reject the conjunction pattern in all cases. In particular, he wants to be able to derive $\bigcirc(p \wedge q)$ from $\bigcirc p \wedge \bigcirc q$ when p and q are distinct propositional atoms. His suggestion is that a restriction should be placed on the conjunction pattern: one derives $\bigcirc(\alpha_1 \wedge \alpha_2)$ from $\bigcirc \alpha_1$ and $\bigcirc \alpha_2$ only if $\alpha_1 \wedge \alpha_2$ is consistent. He calls the latter inference pattern *Consistent Aggregation*, renamed to restricted conjunction (RAND) by van der Torre and Tan in their following variant of van Fraassen's suggestion.

Example 2.1 (Van Fraassen's paradox [29]). *Consider a deontic logic without nested modal operators in which dilemmas like $\bigcirc p \wedge \bigcirc \neg p$ are consistent, but which validates $\neg \bigcirc \bot$, where $\bot$ stands for any contradiction like $p \wedge \neg p$. Moreover, assume that it satisfies replacement of logical equivalents and at least the following two inference patterns Restricted Conjunction (RAND), also called consistent aggregation, and Weakening (W), where $\Diamond \phi$ can be read as "ϕ is possible" (possibility is not necessarily the same as consistency).*

$$\text{RAND:} \frac{\bigcirc \alpha_1, \bigcirc \alpha_2, \Diamond(\alpha_1 \wedge \alpha_2)}{\bigcirc(\alpha_1 \wedge \alpha_2)} \qquad \text{W:} \frac{\bigcirc \alpha_1}{\bigcirc(\alpha_1 \vee \alpha_2)}$$

Moreover, assume the two premises 'Honor thy father or thy mother!' $\bigcirc(f \vee m)$ *and* 'Honor not thy mother!' $\bigcirc \neg m$. *The left derivation of Figure 1 illustrates how the desired conclusion* 'thou shalt honor thy father' $\bigcirc f$ *can be derived from the premises. Unfortunately, the right derivation of Figure 1 illustrates that we cannot accept restricted conjunction and weakening in a monadic deontic logic, because we can derive every $\bigcirc \beta$ from $\bigcirc \alpha$ and $\bigcirc \neg \alpha$.*

$$\frac{\bigcirc(f \vee m) \quad \bigcirc \neg m}{\dfrac{\bigcirc(f \wedge \neg m)}{\bigcirc f} \text{W}} \text{RAND} \qquad \frac{\dfrac{\bigcirc \alpha}{\bigcirc(\alpha \vee \beta)} \text{W} \quad \bigcirc \neg \alpha}{\dfrac{\bigcirc(\neg \alpha \wedge \beta)}{\bigcirc \beta} \text{W}} \text{RAND}$$

Figure 1: Van Fraassen's paradox

Van Fraassen's paradox has a counterpart in dyadic deontic logic. The paradox consists in deriving $\bigcirc(\gamma|\beta)$ from $\bigcirc(\alpha|\beta)$ and $\bigcirc(\neg \alpha|\beta)$ using the following rules of *Restricted Conjunction for the Consequent* (RANDC) and *Weakening of the Consequent* (WC).

$$\text{RANDC}: \frac{\bigcirc(\alpha_1|\beta), \bigcirc(\alpha_2|\beta), \Diamond(\alpha_1 \wedge \alpha_2)}{\bigcirc(\alpha_1 \wedge \alpha_2|\beta)} \qquad \text{WC}: \frac{\bigcirc(\alpha_1|\beta)}{\bigcirc(\alpha_1 \vee \alpha_2|\beta)}$$

2.1.2 Violation Detection Problem: Unrestricted versus Restricted Conjunction

Whereas $p \wedge \neg p$ can not be derived in a paraconsistent logic, we can consistently represent the formula $\bigcirc(p \wedge \neg p)$ in a modal logic, and we can block deontic explosion using a minimal modal logic [3]. This raises the question whether we should accept the conjunction pattern unrestrictedly or in its restricted form.

The choice between the two can be illustrated as follows. Suppose we can derive the obligation $\bigcirc(p \wedge \neg p)$ from $\bigcirc(p)$ and $\bigcirc(\neg p)$ without deriving $\bigcirc f$, or any other counterintuitive consequence. In that case, is $\bigcirc(p \wedge \neg p)$ by itself a consequence we want to block? This presents us with a choice. On the one hand we would like to block $\bigcirc(p \wedge \neg p)$, because it contradicts the "ought implies can" principle. On the other hand, we would like to allow the derivation of $\bigcirc(p \wedge \neg p)$, because such a formula represents explicitly the fact that there is a dilemma.

This choice is even more subtle in dyadic deontic logic. There is the extra question as to whether the "ought implies can" reading implies that the obligation in the consequent must only be consistent in itself, or consistent with the antecedent too. The latter requirement is represented by the following variant of the *Restricted Conjunction for the Consequent* pattern, which we call RANDC2.

$$\text{RANDC2}: \frac{\bigcirc(\alpha_1|\beta), \bigcirc(\alpha_2|\beta), \Diamond(\alpha_1 \wedge \alpha_2 \wedge \beta)}{\bigcirc(\alpha_1 \wedge \alpha_2|\beta)}$$

On the one hand, given $\bigcirc(p|\neg p \vee \neg q)$ and $\bigcirc(q|\neg p \vee \neg q)$, we would like to block the derivation of $\bigcirc(p \wedge q|\neg p \vee \neg q)$ because "ought implies can". On the other hand, we would like to be able to derive it in order to make explicit that $\neg p \vee \neg q$ gives rise to a dilemma, and is not consistent with the fulfillment of the two obligations appearing as premises.

The alternative restricted conjunction pattern RANDC2 highlights the distinction between what we call the violability and the temporal interpretation of dyadic deontic logic. The former interprets the obligation $O(\alpha|\beta)$ as "given that β has been settled beyond repair, we should do α to make the best out of the sad circumstances" [7] and the latter as "if α is the case now, what should be the case next?" The violability interpretation says that $O(\neg \alpha|\alpha)$ represents that α is a violation. For example, if you are going to kill, then do it gently. The temporal interpretation says that the present situation must be changed—which may or may not indicate a violation. For example, the temporal interpretation may be used to express a conditional obligation like "if the light is on, turn it off!"

We would like to point out that the violability interpretation is more expressive, in the sense that the temporal interpretation can be represented by introducing distinct propositional letters for what is the case now, and what is the case in the next moment. For example, "if the light is on, turn it off" can be represented by $\bigcirc(\neg on_2|on_1)$, where on_1 represents that the light is on now, and on_2 that it is on at the next moment in time. In the temporal interpretation, however, it seems impossible to represent all violations in a natural way. Thus, a temporal interpretation with future directed obligations only seems to be a strong limitation.

We use the name "violation detection problem" to refer to the phenomenon that with the restricted conjunction pattern the representation (and hence the detection) of violations is made impossible. We continue the discussion on the violation detection problem in Section 2.2, where we discuss restricted inference patterns formalising contrary-to-duty reasoning.

2.1.3 Forbidden Conflicts

Here is another question raised by dilemmas: is it forbidden to create a dilemma? The following inference pattern is called *Forbidden Conflict* (FC). If the inference pattern is accepted, then it is not allowed to bring about a conflict, because a conflict is sub-ideal.

$$\text{FC}: \frac{\bigcirc(\alpha_1 \wedge \alpha_2 | \beta_1), \bigcirc(\neg \alpha_1 \wedge \alpha_3 | \beta_1 \wedge \beta_2)}{\bigcirc(\neg \beta_2 | \beta_1)}$$

Here is an example, taken from van der Torre and Tan [28]. Assume the premises $\bigcirc k$ and $\bigcirc(p \wedge \neg k | d)$, where k can be read as 'keeping a promise', p as 'preventing a disaster' and d as 'a disaster will occur if nothing is done to prevent it'. (FC) yields $\bigcirc \neg d$. There are situations where this is the right outcome. Consider a person having the obligation to keep a promise to show up at a birthday party. We have $\bigcirc k$, but also $\bigcirc(p \wedge \neg k | d)$. She does not want to go, and so before leaving she does something that might result in a disaster later on, like leaving the coffee machine on. During the party, she leaves and goes home, using her second obligation as an excuse. Nobody will contest that leaving the machine on (on purpose) was a violation already, viz. $\bigcirc \neg d$.

An instance of this inference pattern has been discussed in defeasible deontic logic, and we return to it in Section 2.3.

2.1.4 Resolving Dilemmas

To resolve a conflict between an obligation for p and an obligation for $\neg p$, we need additional information. For example, a total preference order on sets of propositions can resolve all dilemmas by picking the preferred set of obligations among the alternatives of the dilemma, and weaker relations on sets of propositions such as a total pre-order or a partial order leaves some dilemmas unresolved.

The most studied source for a preference order over sets of propositions is a preference order over propositions, which is then lifted to an order on sets of propositions. For example, an ordering on obligations can be derived from an ordering on the authorities who created the obligations, or the moment in time they were created. The level of preference of an obligation may reflect its priority.

Consider three obligations with priority 3, 2 and 1, and a dilemma between the first and the latter two. To represent the priority of an obligation, we write it in the $\bigcirc$ notation. A higher number reflects a higher priority.

$$\{③(p \wedge q), ②\neg p, ①\neg q\}$$

In other words, we can either satisfy the most important obligation $③(p \wedge q)$, or two less important obligations $②\neg p$ and $①\neg q$. Can this dilemma be resolved? There are various well known possibilities in the area of non-monotonic logic. Whether they can be used depends on the origin of the priorities and the application.

The issue of lifting priorities from obligations to sets of them gets more challenging when we consider conditional obligations and deontic detachment, as discussed later on in Section 2.7.

2.2 Forrester's Paradox

We first discuss factual detachment in Forrester's paradox, then the problematic derivation of secondary obligations from primary ones, and finally what we call the violation detection problem for Forrester's paradox.

2.2.1 Factual Detachment versus Conjunction

Forrester's paradox consists of the four sentences 'Smith should not kill Jones,' 'if Smith kills Jones, then he should do it gently,' 'Smith kills Jones', and 'killing someone gently logically implies killing him.' The preference based models of dyadic deontic logic give a natural representation of the two obligations: not killing is preferred to gentle killing, and both are preferred to other forms of killing. However, the following example illustrates that it is less clear how to combine dyadic obligation with factual detachment, deriving unconditional obligations from conditional ones.

Example 2.2 (Forrester's paradox). *Assume a dyadic deontic logic without nested modal operators that has at least replacement of logical equivalents, the Conjunction pattern* AND *and the following inference pattern called factual detachment* FD.

$$\text{FD} : \frac{\bigcirc(\alpha|\beta), \beta}{\bigcirc \alpha}$$

Furthermore, assume the following premise set with background knowledge that gentle murder implies murder $\vdash g \to k$.

$$S = \{\bigcirc(\neg k|\top), \bigcirc(g|k), k\}$$

The set S represents the Forrester paradox when k is read as 'Smith kills Jones' and g as 'Smith kills Jones gently.' We say that the last obligation is a contrary-to-duty obligation with respect to the first obligation, because its antecedent is contradictory with the consequent of the first obligation. Figure 2 visualizes how we can represent the concept of contrary-to-duty as a binary relation among dyadic obligations: the obligation $\bigcirc(\alpha_2|\beta_2)$ is a contrary-to-duty with respect to $\bigcirc(\alpha_1|\beta_1)$ if and only if $\beta_2 \wedge \alpha_1$ is inconsistent.

$$\bigcirc(\neg k|\top)$$
$$\text{inconsistent} \searrow$$
$$\bigcirc(g|k)$$

Figure 2: $\bigcirc(g|k)$ is a contrary-to-duty obligation with respect to $\bigcirc(\neg k|\top)$

The derivation in Figure 3 illustrates how the obligation $\bigcirc(\neg k \wedge g)$, i.e. $\bigcirc(\bot)$, can be derived from S by FD and AND.

$$\frac{\dfrac{\bigcirc(\neg k|\top) \quad \top}{\bigcirc(\neg k)}\text{FD} \quad \dfrac{\bigcirc(g|k) \quad k}{\bigcirc(g)}\text{FD}}{\bigcirc(\neg k \wedge g)}\text{AND}$$

Figure 3: Forrester's paradox

Forrester's paradox can be given two interpretations. First, the dilemma interpretation says that the two obligations give rise to a dilemma, just like the obligations $\bigcirc p$ and $\bigcirc \neg p$ in van Fraassen's paradox. Consequently, according to the dilemma interpretation, there is no problem, the derivation of $\bigcirc(\bot)$ just reflects the fact that there is a dilemma.

The coherent interpretation appeals to the independent and seemingly plausible principle 'ought implies can', $\neg \bigcirc(\bot|\alpha)$. According to this interpretation, the Forrester set is intuitively consistent with the 'ought implies can' principle, and so there is no dilemma, just an obligation to act as good as possible in the sub-ideal situation where the primary obligation has been violated.

There is a consensus in the literature that the example should be given a coherent interpretation, and that the dilemma interpretation is wrong.

2.2.2 Deriving Secondary Obligations from Primary Ones: Strengthening of the Antecedent versus Weakening of the Consequent

The following example shows that Forrester's paradox can be used also to illustrate that combining the desirable inference patterns strengthening of the antecedent and weakening of the consequent is problematic in dyadic deontic logic. For example, strengthening of the antecedent is used to derive 'Smith should not kill Jones in the morning' $O(\neg k | m)$ from the obligation 'Smith should not kill Jones' $O(\neg k | \top)$ and weakening of the consequent is used to derive 'Smith should not kill Jones' $O(\neg k | \top)$ from the obligation 'Smith should drive on the right side of the street and not kill Jones' $O(r \wedge \neg k | \top)$.

Example 2.3 (Forrester's paradox, cont'd [29]). *Assume a dyadic deontic logic without nested modal operators that has at least replacement of logical equivalents and the following inference patterns* Strengthening of the Antecedent *(SA), the* Conjunction pattern for the Consequent *(ANDC) and* Weakening of the Consequent *(WC).*

$$SA : \frac{O(\alpha | \beta_1)}{O(\alpha | \beta_1 \wedge \beta_2)} \quad ANDC : \frac{O(\alpha_1 | \beta), O(\alpha_2 | \beta)}{O(\alpha_1 \wedge \alpha_2 | \beta)} \quad WC : \frac{O(\alpha_1 | \beta)}{O(\alpha_1 \vee \alpha_2 | \beta)}$$

The derivation in Figure 4 illustrates how the obligation $O(\neg k \wedge g | k)$, *i.e.* $O(\bot | k)$, *can be derived from S by* SA *and* ANDC. *Note that the dyadic obligation* $O(\neg k | k)$ *can be given only a violability interpretation in this example, not a temporal interpretation, because it is impossible to undo a killing. That is, this dyadic obligation can be read only as "if Smith kills Jones, then this is a violation."*

$$\frac{\dfrac{O(\neg k | \top)}{O(\neg k | k)} SA \quad O(g|k)}{O(\neg k \wedge g | k)} ANDC \qquad \frac{\dfrac{\dfrac{O(\neg k | \top)}{O(\neg g | \top)} WC}{O(\neg g | k)} RSA \quad O(g|k)}{O(\neg g \wedge g | k)} ANDC$$

Figure 4: Forrester's paradox

The derivation is blocked when SA *is replaced by the following inference pattern* Restricted Strengthening of the Antecedent *(RSA).*

$$RSA : \frac{O(\alpha | \beta_1), \Diamond(\alpha \wedge \beta_1 \wedge \beta_2)}{O(\alpha | \beta_1 \wedge \beta_2)}$$

However, the obligation $O(\bot | k)$ *can still be derived from S by* WC, RSA *and* ANDC. *This derivation from the set of obligations is represented on the right hand side of Figure 4. Like in Example 2.2, we can give the set a dilemma or a coherent interpretation.*

The underlying problem of the counterintuitive derivation in Figure 4 is the derivation of $\bigcirc(\neg g|k)$ from the first premise $\bigcirc(\neg k|\top)$ by WC and RSA, because it derives a contrary-to-duty obligation from its own primary obligation.

Since there is consensus that Forrester's paradox should be given a coherent interpretation, Forrester's paradox in Example 2.3 shows that combining strengthening of the antecedent and weakening of the consequent is problematic for *all* deontic logics.

2.2.3 Violation Detection Problem: Restricted versus Unrestricted Strengthening of the Antecedent

The choice between the unrestricted version and the restricted version of the law of strengthening of the antecedent has some similarity with the choice between the unrestricted version and the restricted version of the law of conjunction. This can be illustrated as follows. Suppose we have the obligation $\bigcirc(\neg k|\top)$. In that case, is $\bigcirc(\neg k|k)$ a consequence we want to block? This presents us with a choice. On the one hand, we would like to block $\bigcirc(\neg k|k)$, because it contradics the "ought implies can" principle. On the other hand, we would like to allow the derivation of $\bigcirc(\neg k|k)$, because this formula represents explicitly that there is a violation. (Cf. our explanatory comments on the violability interpretation, on p. 10.)

The following inference pattern *Violation Detection* (VD) formalizes the intuition that an obligation cannot be defeated by only violating it, and represents a solution to the violation detection problem. The VD pattern models the intuition that after violation the obligation to do α is still in force. Even if you drive too fast, you are still obliged to obey the speed limit.

$$\text{VD} : \frac{\bigcirc(\alpha|\beta)}{\bigcirc(\alpha|\beta \wedge \neg\alpha)} \qquad \text{VD}^- : \frac{\bigcirc(\alpha|\beta \wedge \neg\alpha)}{\bigcirc(\alpha|\beta)}$$

The inverse pattern VD$^-$ says that violations do not come out of the blue. Although this inference pattern may seem intuitive at first sight, it appears too strong on further inspection.

Example 2.4 (Metro). *Consider the following derivation.*

$$\frac{\dfrac{\bigcirc(\alpha|\beta)}{\bigcirc(\alpha|\beta \wedge \neg\alpha)} \text{ VD}}{\bigcirc(\alpha|\alpha \vee \beta)} \text{ VD}^-$$

For example, assume that if you travel by metro, you must have a ticket. We can derive that traveling by metro without a ticket is a violation. The two inference patterns together would derive that if you travel by metro or you buy a ticket, then you must buy a ticket. This is counterintuitive, because buying a ticket without traveling by metro does not involve any obligations. The example illustrates how reasoning about violations only can lead to the wrong conclusions.

Normative systems typically associate sanctions with violations, as an incentive for agents to obey the norms. Such sanctions can sometimes be expressed as contrary-to-duty obligations: the sanction to pay a fine if you do not return the book to the library in time, can be modelled as a contrary-to-duty obligation to pay the fine. By symmetry, though this is less often implemented in normative systems, rewards can be associated with compliance of obligations. In modal logic, an obligation for α is fulfilled if we have $\alpha \wedge \bigcirc \alpha$.

The following inference pattern *Compliance Detection* (CD) formalizes the intuition that an obligation cannot be defeated by only complying with it, analogous to the *Violation Detection* (VD) pattern.

$$\text{CD} : \frac{\bigcirc(\alpha|\beta)}{\bigcirc(\alpha|\beta \wedge \alpha)} \qquad \text{CD}^- : \frac{\bigcirc(\alpha|\beta \wedge \alpha)}{\bigcirc(\alpha|\beta)}$$

The following example illustrates that the inference pattern CD should not be confused with the inverse of CD$^-$, which seems to say that fulfilled obligations do not come out of the blue. Although this inference pattern may seem intuitive at first sight, it is highly counterintuitive on further inspection.

Example 2.5 (Forrester, continued). *Consider the following derivation.*

$$\frac{\dfrac{\bigcirc(\alpha \wedge \beta|\alpha)}{\bigcirc(\alpha \wedge \beta|\alpha \wedge \beta)} \text{CD}}{\bigcirc(\alpha \wedge \beta|\top)} \text{CD}^-$$

You should kill gently, if you kill $\bigcirc(k \wedge g|k)$. Hence, by CD, *you should kill gently, if you kill gently $\bigcirc(k \wedge g|k \wedge g)$ (a fulfilled obligation). However, this does not mean that there is an unconditional obligation to kill gently $\bigcirc(k \wedge g|\top)$. Hence, the inference pattern* CD$^-$ *should not be valid.*

Without the CD pattern, we say that the fulfilled obligation "disappears," analogous to violations. A fulfilled obligation also disappears when we have as an axiom of the logic that $\bigcirc(\alpha|\beta) \leftrightarrow \bigcirc(\alpha \wedge \beta|\beta)$, because in that case $\bigcirc(\alpha \wedge \beta|\beta)$ does not hold because β is compliant with a norm.

2.3 Prakken and Sergot's Cottage Regulations

We first discuss the extension of Forrester's paradox with defeasible obligations, then we return to the violation detection problem, and finally we discuss reinstatement.

2.3.1 Violations and Exceptions

The so-called cottage regulations are introduced by Prakken and Sergot [20] to illustrate the distinction between contrary-to-duty reasoning and defeasible reasoning based on exceptional circumstances. It is an extended version of the Forrester or gentle murderer paradox discussed in Section 2.2. The following example is an alphabetic variant of the original example, because we replaced s, to be read as 'the cottage is by the sea,' by d, to be read as 'there is a dog.' Moreover, as is common, instead of representing background knowledge that w implies f, Prakken and Sergot represent a white fence by $w \wedge f$.

Example 2.6 (Cottage regulations [28]). *Assume a deontic logic that validates at least replacement of logical equivalents and the inference pattern* RSA_C.

$$\text{RSA}_C : \frac{\bigcirc(\alpha|\beta_1), C}{\bigcirc(\alpha|\beta_1 \wedge \beta_2)}$$

C: there is no premise $\bigcirc(\alpha' \mid \beta')$ such that $\beta_1 \wedge \beta_2$ logically implies β', β' logically implies β_1 and not vice versa, α and α' are contradictory and $\alpha \wedge \beta'$ is consistent. [26]

RSA_C formalises a principle of specificity to deal with exceptional circumstances. It is illustrated with Figure 5 (a). Suppose we are given these rules: you ought not to eat with your fingers; if you are served asparagus, you ought to eat with your fingers. One does not want to be able to strengthen the first obligation into: if you are served asparagus, you ought not to eat with your fingers. Such a strengthening is blocked by RSA_C.

Now, assume the obligations

$$S = \{\bigcirc(\neg f|\top), \bigcirc(w \wedge f|f), \bigcirc(w \wedge f|d)\},$$

where f can be read as 'there is a fence around your house,' $w \wedge f$ as 'there is a white fence around your house' and d as 'you have a dog.' Notice that $\bigcirc(w \wedge f|f)$ is a contrary-to-duty obligation with respect to $\bigcirc(\neg f|\top)$ and $\bigcirc(w \wedge f|d)$ is not. If all we know is that there is a fence and a dog ($f \wedge d$), then the first obligation in S is intuitively overridden, and therefore it cannot be violated. Hence, the obligation $\bigcirc(\neg f|f \wedge d)$ should **not** be derivable.

However, if all we know is that there is a fence without a dog (f), then the first obligation in S is intuitively not overridden, and therefore it is violated. Hence, the obligation $\bigcirc(\neg f | f)$ should be derivable.

One should be careful not to treat both $\bigcirc(w \wedge f | f)$ and $\bigcirc(w \wedge f | d)$ as more specific obligations that override the obligation $\bigcirc(\neg f | \top)$: this does not hold for $\bigcirc(w \wedge f | f)$. The latter obligation should be treated as a contrary-to-duty obligation, i.e. as a case of violation. This interference of specificity and contrary-to-duty is represented in Figure 5. This figure should be read as follows. Each arrow is a condition: a two-headed arrow is a consistency check, and a single-headed arrow is a logical implication. For example, the condition C formalizes that an obligation $\bigcirc(\alpha | \beta)$ is overridden by $\bigcirc(\alpha' | \beta')$ if the conclusions are contradictory (a consistency check, the double-headed arrow) and the condition of the overriding obligation is more specific (β' logically implies β). Case (a) represents criteria for overridden defeasibility, and case (b) represents criteria for contrary-to-duty. Case (c) shows that the pair $\bigcirc(\neg f | \top)$ and $\bigcirc(w \wedge f | f)$ can be viewed as overridden defeasibility as well as contrary-to-duty.

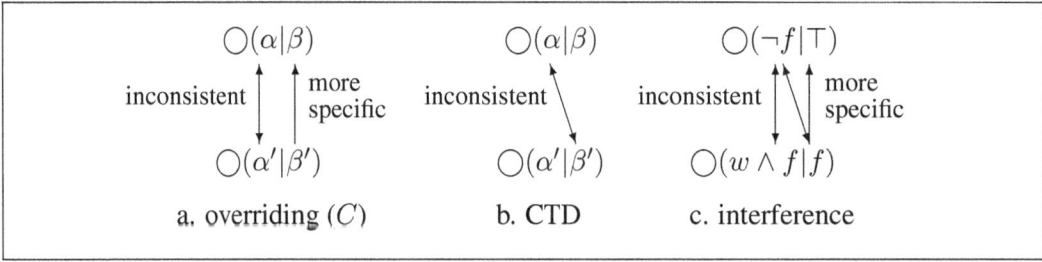

Figure 5: Specificity and CTD

2.3.2 Violation Detection Problem for Defeasible Obligations

What is most striking about the cottage regulations is the observation that when the premise $\bigcirc(\neg f | \top)$ is violated by f, then the obligation for $\neg f$ should be derivable, but not when $\bigcirc(\neg f | \top)$ is overridden by $f \wedge d$. In other words, we have to distinguish violations from exceptions.

In approaches where $\bigcirc(\alpha | \beta)$ implies that $\alpha \wedge \beta$ is consistent, we cannot represent this difference by deriving $\bigcirc(\neg f | f)$ and not deriving $\bigcirc(\neg f | d \wedge f)$. In this sense, this is again an example of the violation detection problem.

We can use priorities to represent the specificity example, by giving the more specific obligation a higher priority. Many conditional logics have specificity built in, but this must be combined with other conflict resolution methods, for example based on time or authority. This is an issue of reasoning about uncertainty, default reasoning, and nonmonotonic logic.

2.3.3 Reinstatement

The question raised by the inference pattern *Reinstatement* (RI) is whether an obligation can be overridden by an overriding obligation that itself is violated. The obligation $\bigcirc(\alpha_1|\beta_1)$ is overridden by $\bigcirc(\neg\alpha_1 \wedge \alpha_2|\beta_1 \wedge \beta_2)$ for $\beta_1 \wedge \beta_2$, but is it also overridden for $\beta_1 \wedge \beta_2 \wedge \neg\alpha_2$? If the last conclusion is not accepted, then the first obligation α_1 should be in force again. Hence, the original obligation is reinstated.

$$\text{RI}: \frac{\bigcirc(\alpha_1|\beta_1), \bigcirc(\neg\alpha_1 \wedge \alpha_2|\beta_1 \wedge \beta_2)}{\bigcirc(\alpha_1|\beta_1 \wedge \beta_2 \wedge \neg\alpha_2)}$$

Suppose you are in the street, and see a child's bike unattended. As a general rule, you should not take the bike, viz. $\bigcirc \neg t$ where t is for taking the bike. Now, suppose you also observe an elderly neighbor collapse with what might be a heart attack. You are a block away from the nearest phone from which you could call for help. In that more specific situation, you should take the bike and go call for help, $\bigcirc(t \wedge h \,|\, e)$, where e and h are for an elderly neighbor collapses and go call for help, respectively. The obligation $\bigcirc \neg t$ is overriden by $\bigcirc(t \wedge h|e)$ for e. But it is not overriden for $e \wedge \neg g$. Of course, if you do not go for help, then the prohibition of t remains.

The following inference pattern RIO is a variant of the previous inference pattern RI, in which the overriding obligation is not factually defeated but overridden. The obligation $\bigcirc(\alpha_1|\beta_1)$ is overridden by $\bigcirc(\neg\alpha_1 \wedge \alpha_2|\beta_1 \wedge \beta_2)$ for $\beta_1 \wedge \beta_2$, and the latter is overridden by $\bigcirc(\neg\alpha_2|\beta_1 \wedge \beta_2 \wedge \beta_3)$ for $\beta_1 \wedge \beta_2 \wedge \beta_3$. The inference pattern RIO says that an obligation cannot be overridden by an obligation that is itself overridden. Hence, an overridden obligation becomes reinstated when its overriding obligation is itself overridden.

$$\text{RIO}: \frac{\bigcirc(\alpha_1|\beta_1), \bigcirc(\neg\alpha_1 \wedge \alpha_2|\beta_1 \wedge \beta_2), \bigcirc(\neg\alpha_2|\beta_1 \wedge \beta_2 \wedge \beta_3)}{\bigcirc(\alpha_1|\beta_1 \wedge \beta_2 \wedge \beta_3)}$$

Example: you should not kill; if you find yourselves in a situation of self-defence, you should kill; if you find yourselves in a situation of self-defence, but your opponent is weak, you should not kill.

Van der Torre and Tan [28] argue that Reinstatement does not hold in general, for example it does not hold for obligations under uncertainty. However, they argue also that these patterns hold for so-called prima facie obligations. The notion of prima facie obligation was introduced by Ross [21]. He writes: 'I suggest *'prima facie* duty' or 'conditional duty' as a brief way of referring to the characteristic (quite distinct from that of being a duty proper) which an act has, in virtue of being of a certain kind (e.g. the keeping of a promise), of being an act which would be a duty proper if it were not at the same time of another kind which is morally significant' [21, p.19]. A prima facie duty is a duty proper when it is not overridden by another prima facie duty. When a prima facie obligation is overridden, it is not a proper duty but it is still in force: 'When we think ourselves justified in breaking, and indeed morally obliged to break, a promise [...] we do not for the moment cease to recognize a prima facie duty to keep our promise' [21, p.28].

Van der Torre and Tan argue also that the inference pattern Forbidden Conflict, discussed in Section 2.1.3, does not hold in general, but it holds for prima facie obligations. If the inference pattern is accepted, then it is not allowed to bring about a conflict, because a conflict is sub-ideal, even when it can be resolved.

2.4 Jeffrey's Disarmament Paradox

In general, reasoning by cases is a desirable property of reasoning with conditionals. In this reasoning scheme, a certain fact is proven by proving it for a set of mutually exclusive and exhaustive circumstances. For example, assume that you want to know whether you want to go to the beach. If you desire to go to the beach when it rains, and you desire to go to the beach when it does not rain, then you may conclude by this scheme 'reasoning by cases' that you desire to go to the beach under all circumstances. The two cases considered here are rain and no rain. This kind of reasoning schemes can be formalized by the following derivation: *If* 'α if β' *and* 'α if not β,' *then* 'α regardless of β.' Formally, if we write the conditional 'α if β' by $\beta > \alpha$, then it is represented by the following disjunction pattern for the antecedent.

$$\text{ORA:} \frac{\beta > \alpha, \neg\beta > \alpha}{\top > \alpha}$$

The following example illustrates that the disjunction pattern for the antecedent combined with strengthening of the antecedent derives counterintuitive consequences in dyadic deontic logic. Example 2.7 is based on the following classic illustration of Jeffrey [11], see also the discussion by Thomason and Horty [25].

Example 2.7 (Disarmament paradox [29]). *Assume a deontic logic that validates at least replacement of logical equivalents and the two inference patterns* RSA *and the* Disjunction pattern for the Antecedent (ORA),

$$\text{ORA}: \frac{\bigcirc(\alpha|\beta_1), \bigcirc(\alpha|\beta_2)}{\bigcirc(\alpha|\beta_1 \vee \beta_2)}$$

and assume as premises the obligations 'we ought to be disarmed if there will be a nuclear war', 'we ought to be disarmed if there will be no war' *and* 'we ought to be armed if we have peace if and only if we are armed'. *They may be formalized as* $\bigcirc(d\,|\,w)$, $\bigcirc(d\,|\,\neg w)$ *and* $\bigcirc(\neg d\,|\,d \leftrightarrow w)$, *respectively. The derivation in Figure 6 shows how we can derive the counterintuitive* $\bigcirc(d \wedge \neg d|d \leftrightarrow w)$. *The derived obligation is inconsistent in most deontic logics, whereas intuitively the set of premises is consistent. The derivation of* $\bigcirc(d|d \leftrightarrow w)$ *is counterintuitive, because it is not possible to fulfill this obligation together with the obligation* $\bigcirc(d\,|\,\neg w)$ *it is derived from. The contradictory fulfillments are respectively* $d \wedge w$ *and* $d \wedge \neg w$.

$$\frac{\dfrac{\dfrac{\bigcirc(d|w) \quad \bigcirc(d|\neg w)}{\bigcirc(d|\top)}\text{ORA}}{\bigcirc(d|d \leftrightarrow w)}\text{RSA} \quad \bigcirc(\neg d|d \leftrightarrow w)}{\bigcirc(d \wedge \neg d|d \leftrightarrow w)}\text{AND}$$

Figure 6: The disarmament paradox

In other words, in this derivation the obligation $\bigcirc(d\,|\,d \leftrightarrow w)$ *is considered to be counterintuitive, because it is not grounded in the premises. If* $d \leftrightarrow w$ *and* w *(the antecedent of the first premise) are true then* d *is trivially true, and if* $d \leftrightarrow w$ *and* $\neg w$ *(the antecedent of the second premise) are true then* d *is trivially false. In other words, if* $d \leftrightarrow w$ *then the first premise cannot be violated and the second premise cannot be fulfilled. Hence, the two premises do not ground the conclusion that for arbitrary* $d \leftrightarrow w$ *we have that* $\neg d$ *is a violation.*

The example is difficult to interpret, because it makes use of a bi-implication. An alternative set of premises, also based on bi-implications, with analogous counterintuitive conclusions is $\{\bigcirc(d|d \leftrightarrow w), \bigcirc(d|\neg d \leftrightarrow w), \bigcirc(\neg d|w)\}$.

ORA also plays a role in the so-called miners' scenario introduced recently by Kolodny and MacFarlane [12].

2.5 Chisholm's Paradox

The second contrary-to-duty paradox we consider is Chisholm [4]'s paradox. We first discuss the choice between deontic versus factual detachment, and then the representation of deontic detachment. We discuss the violation detection problem for deontic detachment only in Section 2.6 after we have introduced Makinson's Möbius strip example.

2.5.1 Deontic versus Factual Detachment

Chisholm's paradox consists of the three obligations of a certain man 'to go to his neighbours assistance,' 'to tell them that he comes if he goes,' and 'not to tell them that he comes if he does not go,' together with the fact 'he does not go.' The preference-based models of dyadic deontic logic again give a natural representation of the three sentences, just like for Forrester's paradox. For example, going to the assistance and telling is preferred to all the other possibilities, and not going to the assistance and not telling is preferred to not going and telling. It seems that the going and not telling and not going and telling may be ordered in various ways. However, the following example illustrates that it is difficult to combine factual with deontic detachment, and to derive unconditional obligations from conditional and unconditional ones.

Example 2.8 (Chisholm's paradox). *Assume a dyadic deontic logic without nested modal operators that has at least replacement of logical equivalents, the Conjunction pattern* AND *factual detachment* FD *and the following inference pattern deontic detachment* DD.

$$\text{DD}: \frac{\bigcirc(\alpha|\beta), \bigcirc\beta}{\bigcirc\alpha}$$

Furthermore, consider the following premise set S.

$$S = \{\bigcirc(a|\top), \bigcirc(t|a), \bigcirc(\neg t|\neg a), \neg a\}$$

The set S formalizes Chisholm's paradox when a is read as 'a certain man goes to the assistance of his neighbors' *and t as* 'the man tells his neighbors that he will come.' *Chisholm's paradox is more complicated than Forrester's paradox, because it also contains an* According-To-Duty (ATD) *obligation. We can represent the notion of according-to-duty as a binary relation among conditional obligations, just like the notion of contrary-to-duty. A conditional obligation $\bigcirc(\alpha \mid \beta)$ is an ATD obligation of $\bigcirc(\alpha_1 \mid \beta_1)$ if and only if β logically implies α_1. The condition of an ATD obligation is satisfied only if the primary obligation is fulfilled. The definition of ATD is analogous to the definition of CTD* *in the sense that an ATD obligation is an obligation conditional upon the fulfilment of an obligation and a CTD obligation is an obligation conditional upon a violation. The second obligation is an ATD obligation and the third obligation is a CTD obligation with respect to the first obligation, see Figure 7.*

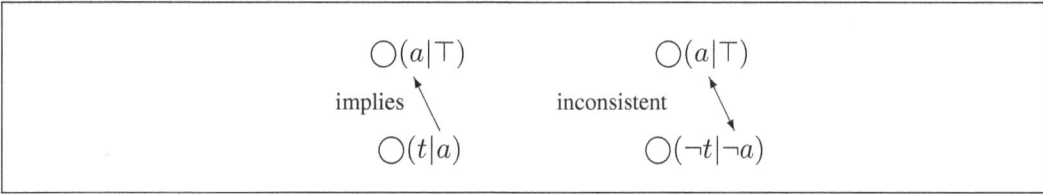

Figure 7: $\bigcirc(t|a)$ is an ATD of $\bigcirc(a|\top)$ and $\bigcirc(\neg t|\neg a)$ is a CTD of $\bigcirc(a|\top)$

The derivation in Figure 8 shows how the counterintuitive obligation $\bigcirc(t \wedge \neg t)$, or $\bigcirc\bot$, can be derived from S by FD, DD *and* AND. *Just like in Forrester's paradox, we can give a dilemma and a coherent interpretation to the scenario, and there is consensus that the latter one is preferred. This is not surprising, as Forrester's paradox shows that factual detachment and conjunction are problematic in themselves.*

$$\cfrac{\bigcirc(t|a) \quad \cfrac{\bigcirc(a|\top) \quad \top}{\bigcirc(a)}\text{FD}}{\bigcirc t}\text{DD} \quad \cfrac{\bigcirc(\neg t|\neg a) \quad \neg a}{\bigcirc(\neg t)}\text{FD}$$
$$\overline{\bigcirc(t \wedge \neg t)}\text{AND}$$

Figure 8: Chisholm's paradox

2.5.2 Deriving Secondary Obligations from Primary Ones: Three Kinds of Transitivity

Deontic detachment is related to the following three variants of transitivity: plain transitivity T, cumulative transitivity CT, and what Parent and van der Torre [18, 19] call aggregative cumulative transitivity ACT.

$$\text{T}: \frac{\bigcirc(\alpha|\beta), \bigcirc(\beta|\gamma)}{\bigcirc(\alpha|\gamma)} \quad \text{CT}: \frac{\bigcirc(\alpha|\beta \wedge \gamma), \bigcirc(\beta|\gamma)}{\bigcirc(\alpha|\gamma)} \quad \text{ACT}: \frac{\bigcirc(\alpha|\beta \wedge \gamma), \bigcirc(\beta|\gamma)}{\bigcirc(\alpha \wedge \beta|\gamma)}$$

The left derivation illustrates that T can be derived from ACT together with SA and WC, and likewise CT can be derived from T and SA, and T can be derived from CT and SA. The right derivation illustrates how ANDC can be derived from SA and ACT. RANDC can be derived analogously from RSA and ACT.

$$\frac{\frac{\bigcirc(\alpha|\beta)}{\bigcirc(\alpha|\beta \wedge \gamma)} \text{SA} \quad \bigcirc(\beta|\gamma)}{\frac{\bigcirc(\alpha \wedge \beta|\gamma)}{\bigcirc(\alpha|\gamma)} \text{WC}} \text{ACT} \qquad \frac{\frac{\bigcirc(\alpha_1|\beta)}{\bigcirc(\alpha_1|\beta \wedge \alpha_2)} \text{SA} \quad \bigcirc(\alpha_2|\beta)}{\bigcirc(\alpha_1 \wedge \alpha_2|\beta)} \text{ACT}$$

The following variant of Chisholm's paradox illustrates that only ACT can be combined with restricted strengthening of the antecedent.

Example 2.9 (Chisholm's paradox, continued). *Assume a dyadic deontic logic that validates at least replacement of logical equivalents and the (intuitively valid) inference patterns RSA (or SA), T (or CT), and ANDC.*

The left derivation in Figure 9 illustrates how the counterintuitive $\bigcirc(\bot | \neg a)$ can be derived from S. Again we can give a dilemma and a coherent interpretation, and there is consensus in the literature that it should get a coherent interpretation. The underlying problem is the derivation of $\bigcirc(t \mid \neg a)$, which seems counterintuitive since it derives a contrary-to-duty obligation from the primary $\bigcirc(a|\top)$. If we accept RSA, then we cannot accept T or CT.

$$\frac{\frac{\bigcirc(t|a) \quad \bigcirc(a|\top)}{\bigcirc(t|\top)} \text{T/CT}}{\bigcirc(t|\neg a)} \text{RSA} \quad \bigcirc(\neg t|\neg a)}{\bigcirc(t \wedge \neg t|\neg a)} \text{AND} \qquad \frac{\frac{\frac{\bigcirc(t|a) \quad \bigcirc(a|\top)}{\bigcirc(a \wedge t|\top)} \text{ACT}}{\bigcirc(t|\top)} \text{WC}}{\bigcirc(t|\neg a)} \text{RSA} \quad \bigcirc(\neg t|\neg a)}{\bigcirc(t \wedge \neg t|\neg a)} \text{AND}$$

Figure 9: Chisholm's paradox

Assume a dyadic deontic logic that validates at least replacement of logical equivalents and the (intuitively valid) inference patterns RSA, ANDC, WC and ACT. The right derivation of Figure 9 illustrates how the counterintuitive $\bigcirc(\bot|\neg a)$ can be derived from S. However, without WC the counterintuitive obligation cannot be derived.

When we compare the two derivations of the contrary-to-duty paradoxes in dyadic deontic logic, we find the following similarity. The underlying problem of the counterintuitive derivations is the derivation of the obligation $\bigcirc(\alpha_1|\neg\alpha_2)$ from $\bigcirc(\alpha_1 \wedge \alpha_2|\top)$ by WC and

RSA. It is respectively the derivation of $\bigcirc(\neg g|k)$ from $\bigcirc(\neg k|\top)$ in Figure 3 and $\bigcirc(t|\neg a)$ from $\bigcirc(a \wedge t|\top)$ in Figure 9. The underlying problem of the contrary-to-duty paradoxes is that a contrary-to-duty obligation can be derived from its primary obligation. It is no surprise that this derivation causes paradoxes. The derivation of a secondary obligation from a primary obligation confuses the different contexts found in contrary-to-duty reasoning. The context of primary obligation is the ideal state, whereas the context of a contrary-to-duty obligation is a violation state. Preference-based deontic logics were developed to semantically distinguish the different violation contexts in a preference ordering, but it appears more challenging to represent these contexts in derivations.

2.6 Makinson's Möbius Strip

Makinson [13]'s Möbius strip illustrates that dilemmas and deontic detachment can also be combined, leading to new challenges and distinctions. We discuss also the violation detection problem for deontic detachment.

2.6.1 Iterated deontic detachment

The so-called Möbius strip (whose name comes from the shape of the example in Figure 10) arises when we allow for deontic detachment to be iterated. We give the version of the example presented by Makinson and van der Torre in their input/output logic, though we use the dyadic representation.

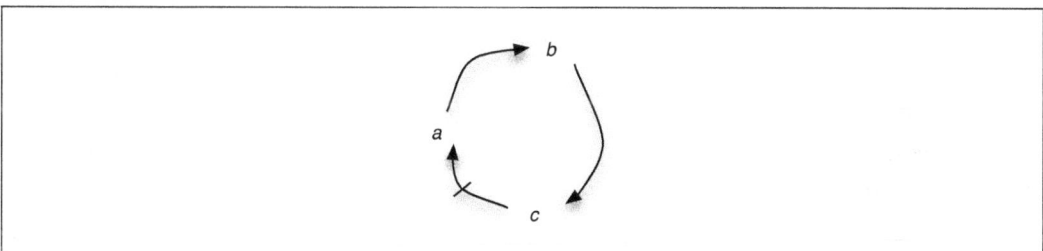

Figure 10: Möbius strip

Example 2.10 (Möbius strip). *Consider three conditional obligations stating $\neg a$ is obligatory given c, that c is obligatory given b, and that b is obligatory given a, together with the* *fact that a is true.*

$$\bigcirc(\neg a|c), \bigcirc(c|b), \bigcirc(b|a), a$$

For instance, a, b, c could represent "Alice (respectively Bob, Carol) is invited to dinner." The obligation $\bigcirc(b|a)$ says that if Alice is invited then Bob should be, and so on.

Makinson [13] gives what we call here the coherent interpretation. He mentions that "intuitively, we would like to have" that under condition a, each of b and c is obligatory, even though we may not want to conclude for $\neg a$ under the same condition. He also indicates that "an approach inspired by maxi choice in AGM theory change" (like the one described in the paper in question) leads to three possible outcomes: both b and c are obligatory; only b is obligatory; neither of b and c is obligatory. The three sets of obligations corresponding to these outcomes are linearly ordered under set-theoretical inclusion.

In their input/output logic framework, Makinson and van der Torre [15] present what we call here the dilemma interpretation of the example. They change the definitions such that precisely the dilemma among these three alternatives is the desired outcome of the example.

There does not seem to be consensus in the literature on which interpretation is the intuitive answer for this example. Deontic detachment has been severely criticised in the literature, so it may be questioned whether full transitivity is natural. However, the choice between coherent and dilemma interpretation is general and can be found in other examples, such as the following variant of Chisholm's paradox.

Example 2.11 (Chisholm's paradox, continued). *Consider this variant of the Möbius strip.*

$$\{\bigcirc(d|c), \bigcirc(c|b), \bigcirc(b|a), a, \neg d\}$$

By symmetry with the dilemma interpretation of Möbius strip, the dilemma interpretation gives three alternatives, $\{\bigcirc b, \bigcirc c\}$, $\{\bigcirc b\}$ and $\emptyset$. Now consider deontic detachment in Chisholm's paradox, together with the fact that we do not tell.

$$\bigcirc(t|a), \bigcirc(a|\top), \neg t$$

Again by symmetry, the dilemma interpretation gives two alternatives, $\{\bigcirc a\}$ and $\emptyset$.

The following example has been introduced by Horty [9] in a prioritised setting, and we will consider it again in the section that comes next. Again the question is raised whether one solution can be a subset of another solution.

Example 2.12 (Order). *Consider the following set of obligations. a is for putting the heating on, and b is for opening the window.*

$$O(a|\top), O(b|\top), O(\neg b|a)$$

The example is a dilemma, but the question is whether there are two or three alternatives. According to the first interpretation, the only two alternatives are the obligations for a and b, and the obligations for a and $\neg b$. According to the second interpretation, there is also the alternative of an obligation for b, without an obligation for a. The latter alternative is a subset of another alternative, analogous to the dilemma interpretation of the Möbius strip example.

2.6.2 Violation detection problem and transitivity

In the previous subsections, like most authors we have assumed that in the Möbius strip the derivation of the obligation for $\neg a$ is intuitively not desirable. However, one can also view it as being intuitively desirable, for the following reason.

Example 2.13 (Möbius strip, continued). *Consider first the coherent interpretation of the Möbius strip, deriving obligations for b and c, but not for $\neg a$. With the transitivity T pattern, one may consider the derivation of the obligation for $\neg a$. This represents that a was actually a violation. With ACT, the violation can be represented by an obligation for $b \wedge c \wedge \neg a$.*

Consider now the dilemma interpretation, presenting three possible outcomes, either $\{Ob, Oc\}$, or $\{Ob\}$, or $\emptyset$. In that case, a leads to a choice, and we may thus have an instance of the forbidden conflict pattern FC that derives that a is forbidden.

2.7 Priority

We are given a set S of conditional obligations along with a priority relation defined on them.

Example 2.14 (Order [9], continued from Example 2.12). *Numbers represent the priority of the obligation, as in Section 2.1.4. Consider*

$$\{③(\neg b|a), ②(b|\top), ①(a|\top)\}$$

①, ②, and ③ can be thought of as expressing commands uttered by a priest, a bishop, and a cardinal, respectively. There are three interpretations. The greedy interpretation derives obligations for a and b. It looks strange, because complying with ①$(a|\top)$ *triggers the most important norm* ③$(\neg b|a)$, *which in turn cancels* ②$(b|\top)$. *To put it another way, complying with* ①$(a|\top)$ *and* ②$(b|\top)$ *results in violating* ③$(\neg b|a)$.

The last link interpretation derives $\bigcirc a$ *and* $\bigcirc \neg b$. *This looks strange too, because* ②$(b|\top)$ *takes precedence over* ①$(a|\top)$, *and* ③$(\neg b|a)$ *will not be triggered (and* ②$(b|\top)$ *cancelled) unless* ①$(a|\top)$ *is fulfilled.*

The weakest link interpretation derives $\bigcirc b$ *only. In order not to trigger* ③$(\neg b|a)$, *and avoid being in a violation state with respect to it, the agent goes for* ②$(b|\top)$ *only.*

The idea underpinning Parent [16]'s next example is similar. Parent argues that different outcomes are expected depending on whether the example is instantiated in the deontic or epistemic domain.

Example 2.15 (Cancer [16]). *Assume we have*

$$\{ ③(c|b), ②(b|a), ①(\neg b|a) \}$$

a is for the set of data used to set up a treatment against cancer, b is for receiving chemo as per the protocol, and c is for keeping WBCs (White Blood Cells) count to a safe level using a drug. In a diagram:

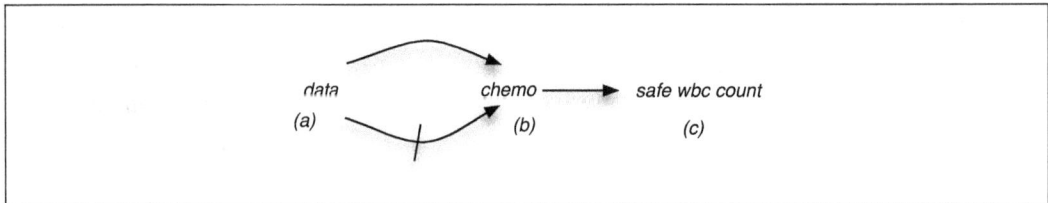

Figure 11: Cancer

Assume the input is a. In that case, we get ②$(b|a)$ and ③$(c|b)$, which derives $\bigcirc b$ and $\bigcirc c$. Given a, both ①$(\neg b|a)$ and ②$(b|a)$ are triggered. These two conflict. The stronger obligation takes precedence over the weaker one.

Assume the input is $\{a, \neg c\}$. In that case, we get ①$(\neg b|a)$ which derives $\bigcirc \neg b$. The reason why may be explained as follows. Following one of Hansson [7]'s suggestions, one might think of the input as someting settled as true. The question is: shall the agent do b or not? The ordering ② > ① says that b has priority over $\neg b$. So it would seem to follow that he should do b. But, in reply, it can be said that the ordering ③ > ② tells us that compliance with the stronger of the two conflicting norms triggers an obligation of even higher rank, namely the obligation to do c. Furthermore, c is already (settled as) false. Hence if the agent goes for b he will put himself in a violation state with respect to a norm with an even higher rank. The only way to avoid the violation of the most important norm is to go for $\neg b$. This is fully in line with what practitioners do: if the WBCs count cannot be maintained at a safe level, chemo is postponed.

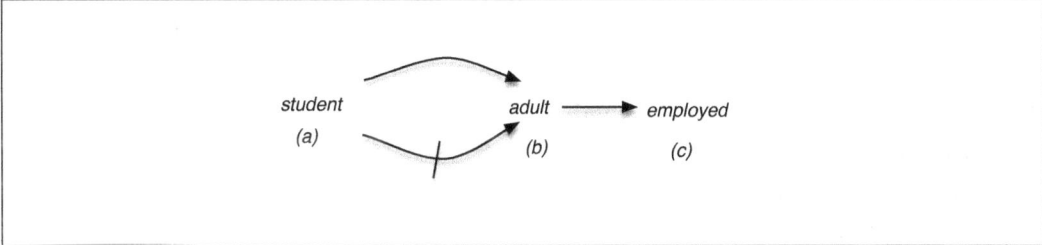

Figure 12: Student example

In the epistemic domain, a different outcome is expected. This can be seen using the reliability interpretation discussed by Horty [9, p. 391] among others. Under the latter interpretation, an epistemic conditional indicates something like a high conditional probability that its conclusion is satisfied, and the priority ordering measures relative strength of these conditional probabilities. For illustration purposes, assume that these conditional probabilities encode statistical assertions about some population groups, and instantiate a, b and c into (this is the example often used to illustrate the non-transitivity of default patterns) *being a student*, *being an adult*, and *being employed*. This is shown in Figure 12. Given input $\{a, \neg c\}$, the expected output remains b.

3 Formal Framework

We extract ten basic properties from the examples, falling in three groups. We believe that the properties of factual detachment and violation detection, the logical properties of substitution, replacement by logical equivalents, implication and paraconsistency are desirable for methods to reason with normative systems, and that the properties of aggregation, factual and norm monotony, and norm induction are optional.

In this section we use the detachment terminology instead of the inference rules terminology.

3.1 Norms, Obligations and Factual Detachment

The distinction between norms and obligations is fundamental in the modern approach to deontic logic. They are related via factual detachment, the detachment of an obligation from a norm.

3.1.1 Representing Norms and Imperatives Explicitly

There are two traditions in normative reasoning, as witnessed by the two historical chapters in the *Handbook on Deontic Logic and Normative Systems* [5]. The first tradition of deontic logic is concerned with logical relations between obligations and permissions, or between the actual and the ideal. The second tradition of normative systems is concerned with normative reasoning, including reasoning about imperatives. Many people suggested a more comprehensive approach, by bringing the two traditions closer to each other, or proposing a uniform approach. For example, when van Fraassen [30] is asking himself whether restricted conjunction can be formalized to reason about dilemmas, he suggests to represent imperatives explicitly.

> "But can this [...] be reflected in the logic of the ought-statements alone? Or can it be expressed only in a language in which we can talk directly about the imperatives as well? This is an important question, because it is the question whether the inferential structure of the 'ought' language game can be stated in so simple a manner that it can be grasped in and by itself. Intuitively, we want to say: there are simple cases, and in the simple cases the axiologist's logic is substantially correct even if it is not in general—but can we state precisely when we find ourselves in such a simple case? These are essentially technical questions for deontic logic, and I shall not pursue them here." [30]

The distinction between norms and obligations was most clearly put forward by Makinson [13], and we follow his notational conventions. To detach an obligation from a norm, there must be a context, and the norms must be conditional. Consequently, norms are a particular kind of rules.

3.1.2 Formal Representation

In this section, a set of norms is represented by a set of pairs of formulae from a base logic, $(a_1, x_1), \ldots, (a_n, x_n)$. A norm (a, x) can be read as "if a is the case, then x ought to be the case." A normative system contains at least one set of norms, the regulative norms from which obligations and prohibitions can be detached. It often contains also permissive norms, from which explicit permissions can be detached, and constitutive norms, from which institutional facts can be detached.

The context is represented by a set of formulae of the same logic. A deontic operator $\bigcirc$ factually detaches obligations, represented by a set of formulae of the base logic, from a set of norms N in a context A, written as $\bigcirc(N, A)$. Unless there is a need for it, we adopt the convention that we do not prefix the detached formula with a modal operator. For example, from a norm that if you travel by metro, you must have a valid ticket $(metro, ticket)$ in the context where you travel by metro, we derive $ticket \in \bigcirc(\{(metro, ticket)\}, \{metro\})$, but $ticket$ itself is not prefixed with a deontic modality. Note that there is no risk of confusing facts and obligations. We know that $ticket$ represents an obligation for $ticket$, because it is factually detached by the $\bigcirc$ operator.

To facilitate presentation and proofs, in this paper we assume propositional logic as the base logic. We write $\beta \in \bigcirc(N, \alpha)$ for $\beta \in \bigcirc(N, \{\alpha\})$, and $\gamma \in \bigcirc((\alpha, \beta), A)$ for $\gamma \in \bigcirc(\{(\alpha, \beta)\}, A)$.

3.1.3 Arguments

Maybe the most important technical innovation of the modern approach is the following convention of writing an argument for α supported by A, traditionally written as $A \therefore \alpha$, as a pair (A, α):

$$(A, \alpha) \in \bigcirc(N) = \alpha \in \bigcirc(N, A)$$

We can move between $\bigcirc(N)$ and $\bigcirc(N, A)$ as we move between $\vdash$ and Cn in classical logic.

It is crucial to understand that the representation of arguments by a pair (A, α) is just a technical method to develop logical machinery: we use it to give more compact representations, to provide proof systems, and to make relations with other branches of logic. However, if you want to know what the argument $(A, \alpha) \in \bigcirc(N)$ *means*, then you always have to translate it back to $\alpha \in \bigcirc(N, A)$.

We reserve the term "norms" to explicit norms, in N. Obviously, one does not derive norms from norms.

In this section we give both the long and the short version of the properties we discuss, to prevent misreading.

3.1.4 Factual Detachment

Factual detachment says that if there is a norm with precisely the context as antecedent, then the output contains the consequent. On the one hand this is relatively weak, as we require the context to be *precisely* the antecedent. A much stronger detachment principle imposes detachment when the antecedent is *implied* by the context. Between these two extremes, we can have that most obligations are detached, or in the most normal cases the obligation is detached. On the other hand the factual detachment principle is also quite strong, as in context a from the norm $(a, \bot)$ the contradiction $\bot$ is detached, and in case of a dilemma of (a, x) and $(a, \neg x)$, in context a both x and $\neg x$ are detached.

Definition 3.1 (Factual detachment). *A deontic operator $\bigcirc$ satisfies the factual detachment property if and only if for all sets of norms N and all sentences α and β we have:*

$$\frac{(\alpha, \beta) \in N}{\beta \in \bigcirc(N, \alpha)}\text{FD} \qquad \frac{(\alpha, \beta) \in N}{(\alpha, \beta) \in \bigcirc(N)}\text{FD} \qquad \frac{(\alpha, \beta) \in N}{(\alpha, \beta)}\text{FD}$$

3.2 Violation Detection

The distinctive feature of norms and obligations with respect to other types of rules and modalities is that they can be violated. Obligations which cannot be violated are not real obligations, but obligations of a degenerated kind. It is not only that ought implies can, but more importantly, ought implies can-be-violated. Issues concerning violations can be found in most deontic examples. For example, dilemma examples arise because some obligation has to be violated, and contrary-to-duty examples arise because some obligation has been violated.

Modal logic offers a simple representation for violations. An obligation for α has been violated if and only we have $\neg \alpha \wedge \bigcirc \alpha$. In our notation with explicit norms, this is $\alpha \in \bigcirc(N, A)$ with $\neg \alpha \in Cn(A)$.

To make sure that violated obligations do not drown, we use the violation detection inference pattern, which we already discussed in Section 2.2.3.

Definition 3.2 (Violation Detection). *A deontic operator $\bigcirc$ satisfies the violation detection property if and only for all sets of norms N, all sets of sentences A and all sentences α we have:*

$$\frac{\alpha \in \bigcirc(N, A)}{\alpha \in \bigcirc(N, A \cup \{\neg \alpha\})}\text{VD} \qquad \frac{(A, \alpha)}{(A \cup \{\neg \alpha\}, \alpha)}\text{VD}$$

Consequently, the restricted strengthening of the antecedent pattern is too weak.

3.3 Substitution

Whereas the first two properties define what is special about *deontic* logic, namely factual detachment and violation detection, the next four properties of substitution, replacements of logical equivalence, implication and paraconsistency say something about *logic*.

The first logical requirement is substitution, well known from classical propositional logic. It says that we can uniformly replace propositional letters by propositional formulae.

Definition 3.3 (Substitution). *Let a uniform substitution map each proposition letter to a propositional formula. A deontic operator $\bigcirc$ satisfies substitution if and only for all sets of norms N, all sets of formulae A, all sentences α and all uniform substitutions σ we have:*

$$\frac{\alpha \in \bigcirc(N, A)}{\alpha[\sigma] \in \bigcirc(N[\sigma], A[\sigma])} \text{SUB}$$

For example, it allows to replace propositional letters by distinct new letters, thus renaming them. This is an example of irrelevance of syntax, a core property of logic.

3.4 Replacement of Logical Equivalents

The following definition introduces two stronger types of irrelevance of syntax.

Definition 3.4 (Irrelevance of Syntax). *Let Cn be closure under logical consequence, and Eq closure under logical equivalence: $\alpha \in Eq(S)$ if and only if there is a β in S such that $Cn(\alpha) = Cn(\beta)$. We write $Eq(a_1, \ldots, a_n)$ for $Eq(\{a_1, \ldots, a_n\})$, and $Cn(a_1, \ldots, a_n)$ for $Cn(\{a_1, \ldots, a_n\})$. Here Cn is the consequence operation of the base logic on top of which the deontic operator $\bigcirc$ operates.*

A deontic operator $\bigcirc$ satisfies formula input (output) irrelevance of syntax if and only for all sets of norms N and all sets of formulae A we have:

$$\bigcirc(N, A) = \bigcirc(N, Eq(A)) \qquad (\bigcirc(N, A) = Eq(\bigcirc(N, A)))$$

and it satisfies set input (output) irrelevance of syntax if and only if for all sets of norms N and all sets of formulae A we have:

$$\bigcirc(N, A) = \bigcirc(N, Cn(A))) \qquad (\bigcirc(N, A) = Cn(\bigcirc(N, A)))$$

The following example illustrates the various types of irrelevance of syntax.

Example 3.5 (Irrelevance of syntax). Let $N = \{(a, x), (a, y)\}$ and $A = \{a\}$. The following table lists some possibilities for $\bigcirc(N, A)$:

$$\emptyset \qquad \{x, y\} \qquad \{x, y, x \wedge y\}$$
$$\{x \wedge y, y \wedge x\} \qquad \{x \wedge y, y \wedge x, x, y\} \qquad \{x \wedge y, y \wedge x, x, y, x \vee y, y \vee x\}$$
$$Eq(x \wedge y) \qquad Eq(x \wedge y, x, y) \qquad Eq(x \wedge y, x, y, x \vee y)$$
$$Cn(x) \cup Cn(y) \qquad Cn(x \wedge y)$$

The first row gives some deontic operators which do not satisfy basic properties. For example, $\emptyset$ does not satisfy factual detachment, $\{x, y\}$ does not satisfy conjunction, and $\{x, y, x \wedge y\}$ does not satisfy variable renaming. That is, if we replace x and y in N, then we end up with the same set, but if we replace x and y in the output, we obtain $y \wedge x$. This violates the most basic property of irrelevance of syntax.

The second row gives some examples satisfying variable renaming for x and y. The set of obligations $\{x \wedge y, y \wedge x\}$ does not satisfy factual detachment again, and the set $\{x \wedge y, y \wedge x, x, y, x \vee y, y \vee x\}$ satisfies besides closure under conjunction also closure under disjunction. Whether this is desired depends on the application. However, all three examples do not satisfy formula output irrelevance of syntax. For example, they all three derive $x \wedge y$, but they do not derive the logically equivalent $x \wedge x \wedge y$.

The third and fourth row close the output under logical equivalence and logical consequence, respectively. $Cn(x \wedge y)$ in the last row satisfies set output irrelevance of syntax.

Input irrelevance is analogous to output irrelevance. For example, when the input is $a \wedge a$ rather than a, it may or may not derive again the same output. If it does not, then the operator violates formula input irrelevance of syntax. Moreover, if it does not treat $\{a, b\}$ and $\{a \wedge b\}$ the same, then it violates input set irrelevance of syntax.

The following example illustrates that output set irrelevance of syntax is too strong in the context of dilemmas, because it may lead to deontic explosion.

Example 3.6 (Irrelevance of syntax, continued). Let

$$N = \{(a, x \wedge y), (a, \neg x \wedge y)\}$$

and $A = \{a\}$. The following table lists some possibilities for $\bigcirc(N, A)$. We only list options closed under logical equivalence, i.e. which satisfy output formula irrelevance of syntax.

$$Eq(x \wedge y, \neg x \wedge y) \qquad Eq(x \wedge y, \neg x \wedge y, x \wedge \neg x \wedge y)$$
$$Eq(x \wedge y, \neg x \wedge y, y) \qquad Eq(x \wedge y, \neg x \wedge y, y, x \wedge \neg x \wedge y)$$
$$Cn(x \wedge y) \cup Cn(\neg x \wedge y) \qquad Cn(x \wedge y) \cup Cn(\neg x \wedge y) \cup Eq(x \wedge \neg x \wedge y)$$
$$Cn(x \wedge y, \neg x \wedge y)$$

The last set $Cn(x \wedge y, \neg x \wedge y)$ derives the whole language, and thus gives rise to explosion. Hence we cannot accept it. The example illustrates that we cannot accept set output irrelevance of syntax.

The difference between the left and right column is that the right column is closed under conjunction, and represents with inconsistent formulae that there is a dilemma.

The difference between the first and the second row is that the second row is closed under disjunction. The difference between the second and the third row is that consistent formulae are closed under logical consequence.

$Cn(x \wedge y) \cup Cn(\neg x \wedge y) \cup Eq(x \wedge \neg x \wedge y))$ has the feature that violations and other obligations are treated in a distinct way.

In this paper we require set input irrelevance of syntax, and formula output irrelevance of syntax. In addition, along the same lines we require that we can replace formulae within the norms by logically equivalent ones. All together, it corresponds to the following property of replacement of logical equivalents.

Definition 3.7 (Replacement of logically equivalent expressions). *We say that two norms ar similar, written as $(\alpha_1, \beta_1) \approx (\alpha_2, \beta_2)$, if and only if $Cn(\alpha_1) = Cn(\alpha_2)$, and $N \approx M$ if and only if for all $(\alpha_1, \beta_1) \in N$ there is a $(\alpha_2, \beta_2) \in M$ such that $(\alpha_1, \beta_1) \approx (\alpha_2, \beta_2)$, and vice versa. A deontic operator $\bigcirc$ satisfies the replacement of Logical Equivalents property if and only if for all sets of norms N and M, all sets of formulae A and B, and all sentences α and β we have:*

$$\frac{N \approx M, Cn(A) = Cn(B), Cn(\alpha) = Cn(\beta), \alpha \in \bigcirc(N, A)}{\beta \in \bigcirc(M, B)} \text{RLE}$$

The examples illustrate that there are other options in between formula and set output irrelevance of syntax, such as requiring that the output is closed under conjunction, or under disjunction, or both. We consider them in Section 3.7.

The principle of irrelevance of syntax has been criticized in belief revision theory. It is discussed by [23] in the context of a study of the notion of revision of a normative system. This notion falls outside the scope of the present paper, and must be left as a topic for future research.

3.5 Implication

The four properties FD, VD, SUB and RLE defined thus far may be called positive properties, in the sense that they require something to be obligatory. That is why we could represent them as Horn rules: given a set of conditions, we require some obligation to be derivable. This contrasts with the examples in Section 2, where typically too much is derived.

The implication requirement in this section and the paraconsistency requirement in the following section may be called negative properties, in the sense that they forbid something to be obligatory. The first requirement makes use of the so-called materialisation of a normative system, which means that each norm (a, x) is interpreted as a material conditional $a \to x$, i.e. as the propositional sentence $\neg a \vee x$. The implication requirement says that if the materializations of N, written as $m(N)$, do not imply $a \to x$, then $(a, x) \notin \bigcirc(N)$. This represents the idea that we cannot derive more than we can derive in propositional logic. In general, implication in the base logic is the upper bound.

Definition 3.8 (Implication). *Let $m(N) = \{a \to x \mid (a, x) \in N\}$ be the set of materializations of N. A deontic operator $\bigcirc$ satisfies the implication property if and only if for all sets of norms N and all sets of sentences A we have $\bigcirc(N, A) \subseteq Cn(m(N) \cup A)$.*

The elements $(\{\alpha\}, \beta)$ of $\bigcirc(N)$ are a subset of $\{(\alpha, \beta) \mid \alpha \to \beta \in Cn(m(N))\}$. In most systems, the base logic is classical propositional logic, but it need not be so. For instance, Cn may be the consequence relation of intuitionistic propositional logic, as in [17]. Cn may also be what Makinson calls a pivotal consequence relation Cn_K, defined by $Cn_K(A) = C(A \cup K)$, where K is a set of formulas, and C is the consequence relation of classical propositional logic. [22] defines and studies two such input/output operations. They are aimed to model the interplay between norms and so-called material dependencies. We have $\bigcirc(N, A) \subseteq Cn_K(m(N) \cup A)$.

3.6 Paraconsistency

To prevent explosion we do not want to derive the whole language, unless maybe in pathological cases in which the normative system contains a norm for each propositional formula. A consequence relation may be said to be paraconsistent if it is not explosive, though there are various ways to make this formal.

To define our paraconsistency requirement, we distinguish obligations representing violations from other obligations. That is, we decompose an operator $\bigcirc(N, A)$ into two operators $V(N, A)$ and $\overline{V}(N, A)$, such that we have $V(N, A) = \{x \in \bigcirc(N, A) \mid \neg x \in Cn(A)\}$ and $\overline{V}(N, A) = \bigcirc(N, A) \setminus V(N, A)$. Trivially, we have

$$\bigcirc(N, A) = V(N, A) \cup \overline{V}(N, A)$$

The basic idea of our paraconsistency requirement is that obligations in $\overline{V}$ can be derived from a set of norms M in N, such that this set of norms M does not explode.

Definition 3.9 (Paraconsistency). *A deontic operator $\bigcirc$ satisfies the paraconsistency property if and only if for all sets of norms N, all sets of formulae A and all sentences α, if $\alpha \in \overline{V}(N, A)$, then there is a $M \subseteq N$ such that $\alpha \in \bigcirc(M, A)$ and $\bigcirc(M, A) \cup A$ is classically consistent.*

Implication and paraconsistency together imply that if $\alpha \in \overline{V}(N, A)$, then there is a $M \subseteq N$ such that $\alpha \in Cn(m(N) \cup A)$ and $\bigcirc(M, A) \cup A$ is classically consistent. This suggest an additional condition: if $\alpha \in \overline{V}(N, A)$, then there is a $M \subseteq N$ such that $\alpha \in Cn(m(N) \cup A)$ and $m(N) \cup A$ is classically consistent.

The underlying intuition to restrict to a set of norms was already raised in Example 1.1 in the introduction. There we observe that if we can derive $\bigcirc(\beta \wedge \gamma)$ from $\bigcirc(\alpha \wedge \beta)$ and $\bigcirc(\neg \alpha \wedge \gamma)$, and we have substitution and replacements of logical equivalents, then we also derive $\bigcirc(\beta)$ from $\bigcirc(\alpha)$ and $\bigcirc(\neg \alpha)$, in other words, we have deontic explosion. This can be verified by replacing β by $\alpha \vee \beta$ and γ by $\neg \alpha \vee \beta$. Therefore, we restrict the set of norms we use to a set of norms which is in some sense "consistent" with the input A.

3.7 Aggregation

The last four properties of aggregation, factual and norm monotony, and norm induction determine the kind of deontic logics we are going to study in our framework. We believe that other choices at this point may be of interest too, but we do not pursue them in this paper.

Aggregation is a core issue in van Fraassen's paradox.

Definition 3.10 (Aggregation). *A deontic operator $\bigcirc$ satisfies the aggregation property if and only if for all sets of norms N, sets of sentences A and sentences α and β we have*

$$\frac{\alpha, \beta \in \bigcirc(N, A)}{\alpha \wedge \beta \in \bigcirc(N, A)} \text{ AND} \qquad \frac{(A, \alpha), (A, \beta)}{(A, \alpha \wedge \beta)} \text{ AND}$$

Van Fraassen's paradox shows that therefore we cannot accept weakening of the consequent. In the context of our present framework, we prefer to call it weakening of the output.

Definition 3.11. *A deontic operator $\bigcirc$ satisfies the weakening of the output property if and only if for all sets of norms N, sets of sentences A and sentences α and β we have*

$$\frac{\alpha \wedge \beta \in \bigcirc(N, A)}{\alpha, \beta \in \bigcirc(N, A)} \text{WO} \qquad \frac{(A, \alpha \wedge \beta)}{(A, \alpha), (A, \beta)} \text{WO}$$

Proposition 3.12. *There is no operator $\bigcirc$ satisfying simultaneously paraconsistency, aggregation, and weakening of the output.*

Proof. Assume the statement does not hold, so there is a deontic $\bigcirc$ satisfying paraconsistency, aggregation and weakening of the output. Consider van Fraassen's paradox $N = \{(\top, p), (\top, \neg p)\}$. According to aggregation and weakening of the output, we have $(\top, q) \in \bigcirc(N)$. According to paraconsistency, $(\top, q) \notin \bigcirc(N)$. Contradiction. □

3.8 Factual Monotony

In this paper we are interested in monotonic logics. Though non-monotonic logics may have their applications too, we believe they should be build on top of the monotonic ones.

Definition 3.13 (Factual monotony). *The factual monotony property holds for $\bigcirc$ if and only if for all sets of norms N, and all sets of sentences A and B, we have $\bigcirc(N, A) \subseteq \bigcirc(N, A \cup B)$.*

As this implies strengthening of the antecedent, Forrester's paradox illustrates that we cannot accept weakening of the consequent.

Proposition 3.14. *There is no operator $\bigcirc$ satisfying simultaneously paraconsistency, factual monotony, and weakening of the output.*

Proof. Assume the statement does not hold, so there is a deontic $\bigcirc$ satisfying paraconsistency, factual monotony and weakening of the output. Consider the first norm of Forrester's paradox $N = \{(\top, \neg k)\}$. According to factual monotony and weakening of the output, we have $(k, \neg k \vee g) \in \bigcirc(N)$. According to paraconsistency, $(k, \neg k \vee g) \notin \bigcirc(N)$. Contradiction. □

3.9 Norm Monotony

Definition 3.15 (Norm monotony). *A deontic operator $\bigcirc$ satisfies the property of norm monotony if and only if for all sets of norms N and M we have $\bigcirc(N) \subseteq \bigcirc(N \cup M)$.*

A deontic operator $\bigcirc$ satisfies the property of monotony if and only if it satisfies those of factual and norm monotony, i.e. for all N, M, A, B we have $\bigcirc(N, A) \subseteq \bigcirc(N \cup M, A \cup B)$.

3.10 Norm Induction

Norm induction says that if there is an output β for an input α, and we add the norm (α, β) to the normative system, then for all inputs, the output of the normative system stays the same. We call it norm induction, because the norm is induced from the relation between facts and obligations. The norm induction requirement considers a set M of such pairs (α, β).

Definition 3.16 (Norm induction). *A deontic operator $\bigcirc$ verifies the property of norm induction if and only if for all sets of norms N and M and all sets of sentences A we have $M \subseteq \bigcirc(N) \Rightarrow \bigcirc(N) = \bigcirc(N \cup M)$*

The strong norm induction principle strengthens the norm induction principle to expansion of the normative system with new norms.

Definition 3.17 (Strong norm induction). *A deontic operator $\bigcirc$ satisfies the property of strong norm induction if and only if for all sets of norms N, N', M, and all sets of sentences A we have $M \subseteq \bigcirc(N) \Rightarrow \bigcirc(N \cup N') = \bigcirc(N \cup N' \cup M)$*

Clearly we have that the strong norm induction property implies the norm induction property.

Together, factual detachment, monotony and norm induction are equivalent to requiring that $\bigcirc$ is a closure operator.

Definition 3.18 (Closure operator). *$\bigcirc$ is a closure operator if and only if it satisfies the following three properties:*

INCLUSION $N \subseteq \bigcirc(N)$

MONOTONY $N \subseteq M$ implies $\bigcirc(N) \subseteq \bigcirc(M)$

IDEMPOTENCE $\bigcirc(N) = \bigcirc(\bigcirc(N))$

Their counterparts in terms of Cn are knowns as the "Tarskian" conditions, after A. Tarski. They can each be rephrased in terms of $\vdash$ ('proves') as follows.

REFLEXIVITY $A \vdash x$ for all $x \in A$

MONOTONY $A \vdash x$ implies $A \cup B \vdash x$

TRANSITIVITY $A \vdash x$ for all $x \in B$ and $B \vdash y$ imply $A \vdash y$

Inclusion for Cn translates into reflexivity of $\vdash$. Monotony for Cn translates into monotony of $\vdash$. Idempotence of Cn corresponds to the transitivity of $\vdash$.

4 Summary

Table 2 lists the examples we discussed in this paper. Given that the world is full of conflicts, we have that normative systems are developed by humans and full of inconsistencies. We need to represent dilemmas consistently, if only to consider their resolution. Van Fraassen's paradox illustrates that doing so presents a basic dilemma: do we accept aggregation or closure under consequence? Forrester's paradox seems to indicate a dilemma too, as it presents two alternatives. In the cottage regulations, such a dilemma interpretation makes sense: either remove the fence, or paint it white. However, in Forrester's gentle murderer example, you cannot undo killing someone. So only the coherent interpretation makes sense. Dilemmas can be resolved by explicit priorities, for example reflecting the authority creating the obligation, or it can be derived from the specificity of the obligations. In the latter case, as illustrated by the cottage regulations, we have to be careful to distinguish violations from exceptions. Jeffrey's disarmament illustrates the problem of reasoning by cases in deontic reasoning. When conditions have an epistemic reading, reasoning by cases may not be valid. Deontic detachment and transitivity originate from Chisholm's paradox, though it is known in the literature as a contrary-to-duty paradox rather than a deontic detachment paradox. Chisholm's paradox illustrates that an alternative representation of the transitivity pattern makes it analogous to Forrester's paradox. Makinson's Möbius strip illustrates many of the problems of reasoning with transitivity. In particular, the dilemma interpretation highlights that we can have solutions being a strict subset of other solutions. More priority examples are introduced in the area of epistemic reasoning, and reasoning with defaults.

Ex.	obligations	patterns	
2.1	Fraassen	$\bigcirc p, \bigcirc \neg p$	AND, WC
2.2	Forrester	$\bigcirc(\neg k\|\top), \bigcirc(g\|k), \vdash g \to k$	FD, (R)AND
2.3	Forrester	$\bigcirc(\neg k\|\top), \bigcirc(g\|k), \vdash g \to k$	(R)SA, ANDC, WC
2.6	Cottage	$\bigcirc(\neg f\|\top), \bigcirc(w \wedge f\|f), \bigcirc(f\|d)$	RSA_o
2.7	Jeffrey	$\bigcirc(d\|w), \bigcirc(d\|\neg w), \bigcirc(\neg d\|d \leftrightarrow w)$	RSA, ORA
2.8	Chisholm	$\bigcirc(a\|\top), \bigcirc(t\|a), \bigcirc(\neg t\|\neg a), \neg a$	AND, FD, DD
2.9	Chisholm	$\bigcirc(a\|\top), \bigcirc(t\|a), \bigcirc(\neg t\|\neg a), \neg a$	T/ CT / ACT, ANDC
2.10	Möbius	$\bigcirc(\neg a\|c), \bigcirc(c\|b), \bigcirc(b\|a), a$	T/ CT
2.14	Priority	$③(\neg b\|a), ②(b\|\top), ①(a\|\top)$	T/ CT

Table 2: Summary of the examples

Maybe the most important technical innovation of our formal framework is the convention of writing an argument for α supported by A as a pair (A, α) with $(A, \alpha) \in \bigcirc(N)$, which means the same as $\alpha \in \bigcirc(N, A)$. We can move between $\bigcirc(N)$ and $\bigcirc(N, A)$ as we move between $\vdash$ and Cn in classical logic.

The ten properties of our formal framework listed in Table 3. We believe that all deontic logics have to satisfy the deontic properties of factual detachment and violation detection, and the logical properties of substitution, replacement by logical equivalents, implication and paraconsistency. Moreover, we discussed the optional properties of aggregation, factual and norm monotony, and norm induction.

FD	$(\alpha, \beta) \in N \Rightarrow \beta \in \bigcirc(N, \alpha)$	Factual detachment
VD	$(A, \beta) \Rightarrow (A \cup \{\neg\beta\}, \beta)$	Violation detection
SUB	$\alpha \in \bigcirc(N, A) \Rightarrow \alpha[\sigma] \in \bigcirc(N[\sigma], A[\sigma])$	Substitution
RLE	$N \approx M, Cn(A) = Cn(B), Cn(\alpha) = Cn(\beta),$ $(A, \alpha) \in \bigcirc(N) \Rightarrow (B, \beta) \in \bigcirc(M)$	Replacement of equivalents
IMP	$\bigcirc(N, A) \subseteq Cn(m(N) \cup A)$	Implication
PC	$\alpha \in \overline{V}(N, A) \Rightarrow \exists M \subseteq N : \alpha \in \bigcirc(M, A)$ and $\bigcirc(M, A) \cup A$ consistent	Paraconsistency
AND	$(A, \alpha)(A, \beta) \Rightarrow (A, \alpha \wedge \beta)$	Conjunction
FM	$(A, \alpha) \Rightarrow (A \cup B, \alpha)$	Factual monotony
NM	$\bigcirc(N) \subseteq \bigcirc(N \cup M)$	Norm monotony
NI	$M \subseteq O(N) \Rightarrow O(N) = O(N \cup M)$	Norm induction

Table 3: Properties

There are two ways to look at the operator $\bigcirc$. First, given a set of norms, it derives sentences from sentences: $\alpha \in \bigcirc_N(A)$. This is the classical way deontic logics considered normative systems: facts go in, obligations go out. Secondly, it derives arguments from norms: $(A, \alpha) \in \bigcirc(N)$. These two views can be used to summarise our properties as follows.

First, the operator in $(A, \alpha) \in \bigcirc(N)$ must be a closure operator, which means that it satisfies factual detachment, norm monotony and norm induction. In addition, it must satisfy substitution and replacement of logical equivalents. Secondly, the operator in $\alpha \in \bigcirc_N(A)$ must satisfy violation detection, implication, paraconsistency, factual monotony, and aggregation.

The properties of norm monotony and norm induction have the effect that our logics will behave classically as Tarskian consequence operators. However, it is important to realise that the closure properties on $\bigcirc(N)$ are not as innocent as they are in other branches of philosophical logic. In particular norm induction is very strong, because it says that every argument (A, α) can itself be used as a norm. This may be true of some branches of case law, but it is probably too strong to be accepted as a universal law for norms. We therefore expect that future studies will first relax this requirement, before relaxing the others.

Finally, we may consider our ten properties as requirements for the further development of reasoning methods for normative systems and deontic logic. We have recently presented two logics satisfying all ten properties [19], which shows that the ten properties are consistent in the sense that they can be satisfied simultaneously.

References

[1] A. R. Anderson. The formal analysis of normative systems. In N. Rescher, editor, *The Logic of Decision and Action*, pages 147–213. Univ. Pittsburgh, 1967, 1956.

[2] L. Åqvist. Deontic logic. In D. Gabbay and F. Guenthner, editors, *Handbook of philosophical logic*, volume 8, pages 147–264. Kluwer Academic publisher, 2002.

[3] B.F. Chellas. *Modal Logic: An Introduction*. Cambridge University Press, 1980.

[4] R.M. Chisholm. Contrary-to-duty imperatives and deontic logic. *Analysis*, 24:33–36, 1963.

[5] D. Gabbay, J. Horty, R. van der Meyden, X. Parent, and L. van der Torre, editors. *Handbook of Deontic Logic and Normative Systems*, volume 1. College Publications, London, UK, 2013.

[6] G. Governatori and M. Hashmi. Permissions in deontic event-calculus. In *Legal Knowledge and Information Systems - JURIX 2015: The Twenty-Elghth Annual Conference, Braga, Portual, December 10-11, 2015*, volume 279, pages 181–182. IOS Press, 2015.

[7] B. Hansson. An analysis of some deontic logics. *Noûs*, 3:373–398, 1969. Reprinted in [8, pp 121-147].

[8] R. Hilpinen, editor. *Deontic Logic: Introductory and Systematic Readings*. Reidel, Dordrecht, 1971.

[9] J. Horty. Defaults with priorities. *Journal of Philosophical Logic*, 36:367–413, 2007.

[10] J. Horty. Deontic modals: why abandon the classical semantics? *Pacific Philosophical Quarterly*, 95:424–460, 2014.

[11] R. Jeffrey. *The Logic of Decision*. University of Chicago Press, 2nd edition, 1983.

[12] N. Kolodny and J. MacFarlane. Iffs and oughts. *Journal of Philosophy*, 107(3):115–143, 2010.

[13] D. Makinson. On a fundamental problem in deontic logic. In P. Mc Namara and H. Prakken, editors, *Norms, Logics and Information Systems*, Frontiers in Artificial Intelligence and Applications, pages 29–54. IOS Press, Amsterdam, 1999.

[14] D. Makinson and L. van der Torre. Input/output logics. *Journal of Philosophical Logic*, 29(4):383–408, 2000.

[15] D. Makinson and L. van der Torre. Constraints for input-output logics. *Journal of Philosophical Logic*, 30(2):155–185, 2001.

[16] X. Parent. Moral particularism in the light of deontic logic. *Artificial Intelligence and Law*, 19(2-3):75–98, 2011.

[17] X. Parent, D. Gabbay, and L. van der Torre. Intuitionistic basis for input/output logic. In S. O. Hansson, editor, *David Makinson on Classical Methods for Non-Classical Problems*, pages 263–286. Springer Netherlands, Dordrecht, 2014.

[18] X. Parent and L. van der Torre. Aggregative deontic detachment for normative reasoning (short paper). In T. Eiter, C. Baral, and G. De Giacomo, editors, *Principles of Knowledge Representation and Reasoning. Proceedings of the 14th International Conference - KR 14*. AAAI Press, 2014.

[19] X. Parent and L. van der Torre. "Sing and dance!": Input/output logics without weakening. In F. Cariani, D. Grossi, J. Meheus, and X. Parent, editors, *Deontic Logic and Normative Systems - 12th International Conference, DEON 2014, Ghent, Belgium, July 12-15, 2014. Proceedings*, volume 8554 of *Lecture Notes in Computer Science*, pages 149–165. Springer, 2014.

[20] H. Prakken and M.J. Sergot. Contrary-to-duty obligations. *Studia Logica*, 57:91–115, 1996.

[21] D. Ross. *The Right and the Good*. Oxford University Press, 1930.

[22] A. Stolpe. Normative consequence: The problem of keeping it whilst giving it up. In G. Governatori and G. Sartor, editors, *Deontic Logic in Computer Science, 10th International Conference, DEON 2010. Proceedings*, volume 6181 of *Lecture Notes in Computer Science*, pages 174–188. Springer, 2008.

[23] A. Stolpe. Norm-system revision: Theory and application. *Artif. Intell. Law*, 18(3):247–283, 2010.

[24] X. Sun and L. van der Torre. Combining constitutive and regulative norms in input/output logic. In F. Cariani, D. Grossi, J. Meheus, and X. Parent, editors, *Deontic Logic and Normative Systems - 12th International Conference, DEON 2014. Proceedings*, volume 8554 of *Lecture Notes in Computer Science*, pages 241–257. Springer, 2014.

[25] R. Thomason and R. Horty. Nondeterministic action and dominance: foundations for planning and qualitative decision. In *Proceedings of the Sixth Conference on Theoretical Aspects of Rationality and Knowledge (TARK'96)*, pages 229–250. Morgan Kaufmann, 1996.

[26] L. van der Torre. Violated obligations in a defeasible deontic logic. In *Proceedings of the Eleventh European Conference on Artificial Intelligence (ECAI'94)*, pages 371–375. John Wiley & Sons, 1994.

[27] L. van der Torre. Contextual deontic logic: Normative agents, violations and independence. *Ann. Math. Artif. Intell.*, 37(1-2):33–63, 2003.

[28] L. van der Torre and Y.-H. Tan. The many faces of defeasibility in defeasible deontic logic. In

D. Nute, editor, *Defeasible Deontic Logic*, pages 79–121. Kluwer, 1997.

[29] L. van der Torre and Y.-H. Tan. Two-phase deontic logic. *Logique et analyse*, 43(171-172):411–456, 2000.

[30] B.C. van Fraassen. Values and the heart command. *Journal of Philosophy*, 70:5–19, 1973.

HANDLING NORMS IN MULTI-AGENT SYSTEM BY MEANS OF FORMAL ARGUMENTATION

CÉLIA DA COSTA PEREIRA, ANDREA G. B. TETTAMANZI, SERENA VILLATA
Université Côte d'Azur, CNRS, I3S, France
`{celia.pereira,andrea.tettamanzi,villata}@unice.fr`

BEISHUI LIAO
Zhejiang University, China
`baiseliao@zju.edu.cn`

ALESSANDRA MALERBA, ANTONINO ROTOLO
University of Bologna, Italy
`alessandramalerba87@gmail.com, antonino.rotolo@unibo.it`

LEENDERT VAN DER TORRE
University of Luxembourg
`leon.vandertorre@uni.lu`

Abstract

Formal argumentation is used to enrich and analyse normative multi-agent systems in various ways. In this chapter, we discuss three examples from the literature of handling norms by means of formal argumentation. First, we discuss how existing ways to resolve conflicts among norms using priorities can be represented in formal argumentation, by showing that the so-called Greedy and Reduction approaches can be represented using the weakest and the last link principles respectively. Based on such representation results, formal argumentation can be used to explain the detachment of obligations and permissions from hierarchical normative systems in a new way. Second, we discuss how formal argumentation can be used as a general theory for developing new approaches for normative reasoning, using a dynamic ASPIC-based legal argumentation

This work is supported by the European Union's Horizon 2020 research and innovation programme under the Marie Curie grant agreement No: 690974 (Mining and Reasoning with Legal Texts, MIREL).

theory. We show how existing logics of normative systems can be used to analyse such new argumentation systems. Third, we show how argumentation can be used to reason about other challenges in the area of normative multiagent systems as well, by discussing a model for arguing about legal interpretation. In particular, we show how fuzzy logic combined with formal argumentation can be used to reason about the adoption of graded categories and thus address the problem of open texture in normative interpretation. Our aim to discuss these three examples is to inspire new applications of formal argumentation to the challenges of normative reasoning in multiagent systems.

1 Introduction

Norms regulate our everyday life, and are used to assess conformance of behaviour with respect to regulations holding in multi-agent systems. Agents undertake discussions about norms to assess their validity or applicability subject to particular conditions, to derive the obligations and permissions to be enforced, or to claim that a certain normative conclusion cannot be derived from the existing regulations. Given the profound importance of norms in multi-agent systems, it is fundamental to understand, e.g., which norms are valid in certain environments, how to interpret them, and to determine the deontic conclusions of such norms. Some influential philosophers, such as Scott Shapiro [54], argue that the law has an inherent teleological nature and that norms are plans, and in most existing normative multiagent systems, norms are like plans which aim at achieving the social goals the members of a society have decided to share [12, 13]. However, it is not obvious that, for example, norms stating human rights can be considered as plans, and we therefore do not commit here to such philosophical claims.

Formal argumentation is typically based on logical arguments constructed from prioritised rules, and it is no surprise that the first applications of formal argumentation in the area of normative multiagent systems were concerned with the resolution of conflicting norms and norm compliance. Moreover, several frameworks have been proposed for normative and legal argumentation [10], but no comprehensive formal model of normative reasoning from arguments has been proposed yet. In this chapter we discuss three challenges to illustrate the variety of applications of formal argumentation techniques in the field of normative multi-agent systems.

- How can formal argumentation be used to explain existing approaches for reasoning about normative multi-agent systems?

- How can new argumentation systems for reasoning about norms be developed, and how can these new argumentation systems be analysed?

- Which issues in the area of normative multiagent systems can be modelled and analysed using formal argumentation, besides the resolution of conflicting norms and checking compliance of a system with a set of norms?

First, we discuss how existing detachment procedures for prioritized norms can be represented in argumentation, by showing how the so-called Greedy and Reduction approaches can be represented in argumentation by applying the weakest link and the last link principles respectively [35]. Based on such representation results, formal argumentation can be used to explain the detachment of obligations and permissions from hierarchical normative systems in a new way.

Second, we discuss an instance of ASPIC$^+$ [40, 48, 59] capturing the inference schemes of arguments about norms like *legislative and interpretative* arguments. Moreover, we show how to adopt the input/output logic methodology [38] for the analysis of these new argumentation systems [59].

Third, we discuss the model of da Costa Pereira *et al.* [17], in which norm interpretation is a mechanism to deal with uncertainty, in contrast to existing models of norm interpretation in the context of Normative Multi-Agent Systems and AI&Law [12, 13, 66, 4, 39, 5]. This uncertainty reflects that, in legal theory, a definition of an empirical concept bounded in all now-foreseeable dimensions can break down in the face of unforeseen and unforeseeable events, and norms cannot anticipate all potential occurrences falling within the application scope of any legal norm [29, 37]. In other words, it reflects that the interpretation of legal rules is often uncertain: legal language is vague, the concepts used to describe a legal rule are not always precise, and the purpose of the rule may be differently perceived [30, 19, 36]. The model uses fuzzy logic to measure the uncertainty of legal concepts, and argumentation is used to handle the conflicts between different interpretations of norms. More precisely, a fuzzy argumentation system [56] to represent the interpretations, is combined with fuzzy labeling to evaluate the status of fuzzy arguments [18]. As in many logical analyses of legal reasoning, the model is not purely descriptive and it is rather meant to offer a rational reconstruction for explaining and checking the robustness of interpretive arguments. A formal model for legal impreciseness must be cognitively sound, in the sense that it works on reliable cognitive assumptions.

The remainder of the chapter is organised as follows. Second 2 introduces how prioritized norms can be represented in argumentation. In Section 3, we discuss the logical properties of the static legal argumentation system proposed by Prakken and Sartor, and we reformulate it in a normative perspective. Section 4 motivates our adoption of graded categories as a tool to tackle the problem of open texture in legal interpretation. Section 5 introduces a model of fuzzy argumentation and fuzzy labeling, and Section 6 interprets a norm with flexibiity and conducts a case study by using an example from medically assisted reproduction. Second 7 discusses related work and Section 8 concludes.

2 Argumentation semantics for hierarchical normative systems

Consider the following benchmark example introduced by Hansen [28], which we call here the *prioritised triangle* due to its graphical visualization in Figure 1.

Example 1 (Prioritised triangle [28]). *Imagine you have been invited to a party. Before the event, you receive several imperatives, which we consider as the following set of norms.*
- *Your mother says: if you drink (p), then don't drive ($\neg x$).*
- *Your best friend says: if you go to the party (a), then you'll drive (x) us.*
- *An acquaintance says: if you go to the party (a), then have a drink with me (p).*

We assign numerical priorities to these norms, namely '3', '2' and '1' corresponding to the sources 'your mother', 'your best friend' and 'your acquaintance', respectively.

Let a, p and x respectively denote the propositions that you go to the party; you drink; and you drive. In terms of a hierarchical normative systems [1], these norms are respectively represented as $(a,p)_1$, $(p,\neg x)_3$ and $(a,x)_2$. These three norms are visualized in Figure 1(a).

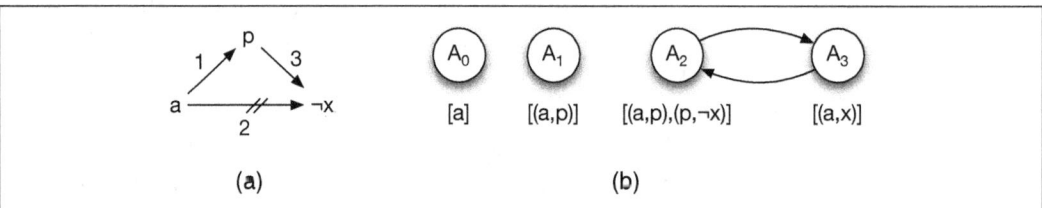

Figure 1: The prioritised normative system of the prioritised triangle example.

Consider the following two approaches resulting in different outcomes or *extensions* [15, 64, 35].

Greedy approach Based on the context, a set of propositions that are known to hold, this approach always applies the norm with the highest priority that does not introduce inconsistency to an extension and the context. Here we say that a norm is applicable when its body is in the context or has been produced by other norms and added to the extension. In this example, we begin with the context $\{a\}$, and (a, x) is first applied. Then (a, p) is applied. Finally, $(p, \neg x)$ cannot be applied as this would result in a conflict, and so, by using the Greedy approach, we obtain the extension $\{p, x\}$.

Reduction approach In this approach, a candidate extension is identified. All norms which are applicable according to this candidate extension are selected and transformed into unconditional or body-free norms (i.e., a norm (a, b) selected in this way is transformed to a norm $(\top, b)$). The modified normative system, with the transformed norms is evaluated using the Greedy approach. The candidate extension is selected as an extension by the Reduction approach if it is identified as an extension according to this application of the Greedy approach. In this example, selecting a candidate extension $\{p, \neg x\}$, we get a set of body-free norms $\{(\top, p), (\top, \neg x), (\top, x)\}$. The priorities assigned to these norms are carried through from the original normative system, and are therefore respectively 1, 3 and 2. After applying the Greedy approach, we get $\{p, \neg x\}$, which is thus an extension of the Reduction approach. If on the other hand we had selected the candidate extension $\{p, x\}$, this new extension would not appear in the greedy evaluation, because $(\top, x)$ has a lower priority than $(\top, \neg x)$. Consequently $\{p, x\}$ is *not* an extension of the Reduction approach.

We now consider the prioritised triangle example in formal argumentation. Given a normative system, we may construct an argumentation framework as illustrated in Figure 1(b), which is a directed graph in which nodes denote arguments, and edges denote attacks between arguments. An argument is represented as a path of a directed graph starting from a node in the context. In this simple example, there are four arguments A_0, A_1, A_2 and A_3, represented as $[a], [a, p], [a, p, \neg x]$ and $[a, x]$, respectively. Since the conclusions of A_2 and A_3 are inconsistent, A_2 attacks A_3 and vice versa. Priorities allow us to transform these attacks into *defeats* according to different principles.

Last link ranks an argument based on the strength of its last inference, if the last link principle is applied, then $[a, p, \neg x]$ defeats $[a, x]$. As result, the principle allows us to conclude $\{p, \neg x\}$.

Weakest link ranks an argument based on the strength of its weakest inference. If the weakest link principle is used instead, $[a, x]$ defeats $[a, p, \neg x]$, and concludes $\{p, x\}$.

In this example, the last link principle thus gives the same result as the Reduction approach, and weakest link gives the same result as the Greedy approach. Liao *et al.* [35] show that this is not a coincidence, but it holds for all totally ordered normative systems. This result addresses the challenge raised by Dung [20] aiming at representing nonmonotonic logics through formal argumentation. In particular, argumentation is a way to exchange and communicate viewpoints, thus having an argumentation theory representing a nonmonotonic logic is desirable for such a logic, in particular when the argumentation theory is simple and efficient. Note that it is not helpful for the development of nonmonotonic logics themselves, but it helps when we want to apply such logics in distributed and multiagent scenarios.

Based on such representation results, formal argumentation can be used to explain the detachment of obligations and permissions from hierarchical normative systems in a new way. Moreover, many other challenges in normative reasoning have been expressed as inconsistent sets of formulas that are intuitively consistent, traditionally called deontic paradoxes. The most well known are the so-called contrary-to-duty paradoxes, which are concerned with handling norm violations. Techniques from non-monotonic reasoning have been applied to handle contrary-to-duty reasoning, and formal argumentation techniques can be applied in the same way [44]. Finally, the most discussed practical problem in normative systems is norm conformance and compliance, which is a computational problem to check whether a business process is in accordance with a set of norms. Handling priorities among norms is again a central challenge for norm compliance, and formal argumentation techniques for resolving conflicts between norms can be extended with reasoning about business processes to reason about norm compliance [57].

3 New argumentation systems for normative reasoning

In the previous section, we used an argumentation system to explain the conclusions that are detached from a hierarchical normative system. The converse is done as well: new argumentation systems for normative reasoning have been developed for normative reasoning, for which detachment procedures have been defined to analyse these argumentation systems. We illustrate this by the argumentation system for legal reasoning proposed by Prakken and Sartor [48], which has been analyzed and extended by van der Torre and Villata [59].

Definition 1 (LAS-PS). *A legal argumentation system or LAS is a tuple $\langle \mathcal{L}, -, \mathcal{R} \rangle$ where $\mathcal{L}$ is the legal language of all sentences α, $- : \mathcal{L} \to 2^{\mathcal{L}}$ is a function given by $-(P) = \{\neg P\}$, $-(\neg P) = \{P\}$ and $-(N) = \emptyset$, and $\mathcal{R}$ contains the Defeasible modus ponens (DMP), rule for each possible norm N of the form $\phi_1 \wedge \ldots \wedge \phi_n \rightsquigarrow \psi$.*
   DMP: $\phi_1, \ldots, \phi_n, \phi_1 \wedge \ldots \wedge \phi_n \rightsquigarrow \psi \Rightarrow \psi$;

In order to illustrate the legal argumentation framework, the running example proposed by Prakken and Sartor [48] is adapted.

Example 2 (Smoking regulations). *Consider propositional atoms $P ::= a|b|c|d|e|f$ where a: "people want to smoke in a closed space", b: "the public place has special* secluded smoking areas", c: "people need to smoke cannabis on medical grounds", d: "people are forbidden from smoking cannabis and tobacco in public places", e: "cannabis is allowed for medical treatment", f: "people are permitted to smoke cannabis in recreational cannabis establishments". $\mathcal{R}$ contains expressions for inference rules DMP of the form, for example: $a, a \leadsto b \Rightarrow b$ and $a, \neg c, a \land \neg c \leadsto d \Rightarrow d$.

Prakken and Sartor [48] follow Modgil and Prakken [40], and do not consider a model theoretic semantics for this language. Instead, they define a set of arguments.

Definition 2 (LAS PS arguments). *A knowledge base K is a set of sentences of $\mathcal{L}$. The set of arguments A on the basis of a knowledge base K in a legal argumentation system LAS is called $Arg(LAS, K)$ and is the smallest set of expressions containing the literals in K and closed under the following rule:*

if $A_1, \ldots, A_n \subseteq Arg(LAS, K)$ and $concl(A_1), \ldots, concl(A_n) \Rightarrow L \in \mathcal{R}$ then we have also $(A_1, \ldots, A_n \Rightarrow L) \in Arg(LAS, K)$,

where $concl(A)$ is defined by $concl(L) = L$ and $concl(A_1, \ldots, A_n \Rightarrow L) = L$. We may leave out the brackets if there is no risk of confusion.

To study this notion of norm based argument, consequence is defined by considering only the conclusions of the arguments, in other words, by abstracting away the explicit arguments. Following input/output logic conventions, the consequence is called *Out*.

Definition 3 (Output PS). $Out(LAS, K) = \{concl(A) \mid A \in Arg(LAS, K)\}$.

Example 3 (Continued). *Consider the knowledge base of the smoking regulations $K_1 = \{a, b, c, e, a \land b \leadsto \neg d, c \land \neg d \land e \leadsto f\}$ where the norms state that*

- *if people want to smoke in a closed space and the public place has smoking special secluded areas, then people are not forbidden from smoking cannabis and tobacco in public places;*

- *if people need to smoke cannabis on medical grounds and it is not forbidden from smoking cannabis and tobacco in public places and cannabis is allowed for medical treatment, then people are permitted to smoke cannabis in recreational cannabis establishments;*

Arguments can be constructed combining DMP inference rules as follows:

- $A_1 : a, b, a \land b \leadsto \neg d \Rightarrow \neg d;$

- $A_2 : c, (a, b, a \land b \leadsto \neg d \Rightarrow \neg d), e, c \land \neg d \land e \leadsto f \Rightarrow f.$

Therefore, from arguments A_1, A_2, we have that $concl(A_1) = \neg d$ and $concl(A_2) = f$. We conclude that $Out(LAS_1, K_1) = \{a, b, c, \neg d, e, f\}$.

We now introduce a logical analysis. Van der Torre and Villata [59] use a proof system with expressions $K \therefore L$. The proof system contains four rules, called Identity (ID), Strengthening of the input (SI), Factual Detachment (FD), and Deontic Detachment (DD). The former is sometimes called Monotonicity (Mon), and the latter two are sometimes called Modus Ponens (MP) or Cumulative Transitivity (CT). The notion of consequence is called simple-minded reusable throughput or out_3^+ by Makinson and van der Torre [38].

Definition 4 (Derivations PS). *$der(LAS)$ is the smallest set of expressions $K \therefore L$ closed under the following four rules.*

ID: $\{L\} \therefore L$ *for a literal L*

SI: *from $K \therefore L$ derive $K \cup K' \therefore L$*

FD: $\{L_1, \ldots, L_n, L_1 \wedge \ldots \wedge L_n \leadsto L\} \therefore L$ *for a norm $L_1 \wedge \ldots \wedge L_n \leadsto L$*

DD: *from $K \therefore L_i$ for $1 \leq i \leq n$ and $K \cup \{L_1 \ldots, L_n\} \therefore L$ derive $K \therefore L$*

The close relation between arguments and derivations in a deontic logic or a logic of normative systems is illustrated by the following property:

$K \therefore L \in der(LAS)$ iff $L \in Out(LAS, K)$.

This is not surprising, as the similarity is quite clear from the structure of arguments. However, making the relation precise by framing the legal argument system into an input/output logic highlights a drawback of the legal argumentation system of Prakken and Sartor: simple-minded reusable throughput is usually adopted for default logics and logic programs, not for the normative reasoning.

To establish the results with constrained input/output logic, only rebut is considered. Thus undercut is not considered. Moreover, they do not consider defeasible knowledge and undermining. So the only attack is the attack of an argument with an opposite literal. This is obviously a very simple notion of attack which is of little use in most applications, but it useful to establish the relation with logical approaches.

Definition 5 (Attack PS). *The set of sub-arguments of B is the smallest set containing B, and closed under the rule: if $A_1, \ldots, A_n \Rightarrow L$ is a sub-argument of B, then $A_1, \ldots, A_n$ are also sub-arguments of B.*

A attacks B iff there is a sub-argument B' of B such that $concl(A) \in -(concl(B'))$. We write $attack(AS, K)$ for the set of all attacks among $Arg(AS, K)$.

A semantics associates sets of extensions with an argumentation framework, where each extension consists of a set of arguments. For each extension, the output consists of the set of conclusions of the arguments, as for *Out* before. A semantics thus gives us a set of sets of conclusions, which is called an *Outfamily*.

Definition 6 (Outfamily PS). *An extension is a set of arguments, and an argumentation semantics $sem(arg, attack)$ is a function that takes as input a set of arguments and a binary attack relation among the arguments, and as output a set of extensions.*

$$Outfamily(K, sem) = \{\{concl(A) \mid A \in S\} \mid S \in sem(arg(AS, K), attack(AS, K))\}.$$

Constrained output in the input/output logic framework is defined as follows, being inspired by maximal consistent set constructions in belief revision and non-monotonic reasoning. *Maxf* takes the maximal sets of norms of K such that the output of K is consistent, and *Outf* takes the output of these maximal norm sets.

Definition 7 (Outf). *Let $K = K^L \cup K^N$ consist of literals K^L and norms K^N.*
$Conf(K) = \{N \subseteq K^N \mid Out(K^L \cup N) \text{ consistent}\}$
$Maxf(K) = \{N \subseteq K^N \mid N \text{ maximal w.r.t } \subseteq \text{ in } Conf(K)\}$
$Outf(K) = \{Out(K^L \cup N) \mid N \in Maxf(K)\}$

Theorem 1 (Characterization PS). *$Outfamily(KB, sem) = Outf(K)$ for sem is stable or preferred.*

Van der Torre and Villata [59] add an additional modal operator O to the language. All norms are of the form $L_1 \wedge \ldots \wedge L_n \rightsquigarrow L$, as before, or $L_1 \wedge \ldots \wedge L_n \rightsquigarrow OL$. The body contains simple literals and the head contains either a literal or an obligation. They redefine the concepts or *LAS*, *Out*, *der*, etc. As there is no risk for confusion, we refer to them with the same names as in the previous sections.

Definition 8 (LAS O). *Given a set of propositional atoms. The literals, norms and legal language $\mathcal{L}$ are given by the following BNF.*
$L ::= P \mid \neg P$ with P in propositional atoms
$M ::= L \mid OL$
$N ::= L \wedge \ldots \wedge L \rightsquigarrow M$
$\alpha ::= L \mid N$
A legal argumentation system with obligations or LAS is as defined before, where the $-$ function is extended to obligations.

The definition of arguments is adapted in the obvious way. In the output, they consider only the obligatory propositions.

Definition 9 (Output O). $Out(LAS, K) = \{L \mid A \in Arg(LAS, K), concl(A) = OL\}$.

Example 4. *We consider a revised version of the running example about smoking regulations. We have that the LAS_2 is based on propositional atoms $P ::= a|b|c|d|e$ where a: "the person wants to smoke in a closed space", b: "the person is in a private space", c: "the person needs to smoke on medical grounds", d: "the person is forbidden from smoking", e: "use electronic cigarettes". and $\mathcal{R}$ contains expressions for inference rules of the form:*

- $a, a \rightsquigarrow b \Rightarrow b$;

- $a, \neg c, a \wedge \neg c \rightsquigarrow d \Rightarrow Od$;

Consider now the extended knowledge base of the smoking regulations represented by $K_2 = \{a, \neg b, \neg c, a \wedge \neg c \rightsquigarrow d, a \wedge b \rightsquigarrow \neg d, c \rightsquigarrow \neg d, a \wedge d \rightsquigarrow Oe\}$ where the norms state that

- *if the person is in a closed space and she does not need to smoke on medical grounds, then the person is forbidden from smoking;*

- *if the person wants to smoke in a closed space and she is in a private space, then the person is not forbidden from smoking;*

- *if the person needs to smoke on medical grounds, then she is not forbidden from smoking;*

- *if the person wants to smoke in a closed space and she is forbidden from smoking, then it is obligatory to use electronic cigarettes;*

We can construct the following arguments:

- $A_1 : a, \neg c, a \wedge \neg c \rightsquigarrow d \Rightarrow d$;

- $A_2 : a, (a, \neg c, a \wedge \neg c \rightsquigarrow d \Rightarrow d), a \wedge d \rightsquigarrow Oe \Rightarrow Oe$;

We have that $concl(A_1) = \{d\}$ and $concl(A_2) = \{Oe\}$, and we can thus conclude $Out(LAS_2, K_2) = \{e\}$ i.e., the conclusion is an obligation to use electronic cigarettes.

The constrained version can be defined analogously.

The proof system contains two rules, Strengthening of the Input (SI) and Factual Detachment (FD). The notion of consequence is called simple-minded output or out_1 by Makinson and van der Torre [38].

Definition 10 (Derivations O). $der(LAS)$ is the smallest set of expressions $K \therefore L$ closed under the following two rules.

SI: from $K \therefore L$ derive $K \cup K' \therefore L$

FD: $\{L_1, \ldots, L_n, L_1 \wedge \ldots \wedge L_n \leadsto L\} \therefore L$ for a norm $L_1 \wedge \ldots \wedge L_n \leadsto OL$

Again we have $K \therefore L \in der(LAS)$ iff $L \in Out(LAS, K)$. The system does not satisfy deontic detachment, e.g. from $K = \{a, a \leadsto Ob, b \leadsto Oc\}$ we cannot derive Oc. This is reflected in the proof system by the lack of the DD rule.

Finally, van der Torre and Villata show how can to redefine the concepts of LAS, Out, der, etc., to re-introducing deontic detachment. This illustrates how the formal analysis can inspire the development of new argumentation systems.

4 From Open Texture to Graded Categories

4.1 Flexible legal interpretation based on graded categories

Legal systems are the product of human mind and are written in natural language. This implies that the basic processes of human cognition have to be taken into account when interpreting norms, and that, as natural languages are inherently vague and imprecise, so are norms.

The application of laws to a new situation is a metaphorical process: the new situation is mapped on to a situation in which applying law is obvious, by analogy. Here, by metaphor we mean using a well understood, prototypical situation to represent and reason about a less understood, novel situation. Metaphors are one of the basic building blocks of human cognition [34].

Norms are written with references to categories. As pointed out by Lakoff [33], "Categorization is not a matter to be taken lightly. There is nothing more basic than categorization to our thought, perception, action, and speech." The "classical theory" that categories are defined by common properties is not entirely wrong, but it is only a small part of the story. It is now clear that categories may be based on prototypes. Some categories are vague or imprecise; some do not have gradation of membership, while others do. The category "US Senator" is well defined, but categories like "rich person" or "tall man" are graded, simply because there are different degrees of richness and tallness. However, it is important to notice that these degrees of membership depend both on the the context in which the norm will be applied and on the goal associated to the norm. To be considered tall in the Netherlands is not the same as to be considered tall in Portugal, for example. We have thus first to consider the context and then the goal associated to the norm.

We explore the use of fuzzy logic as a suitable technical tool to capture the imprecision related to categories. More precisely, a category may be represented as a fuzzy set: the membership of an element to a category is a graded notion.

As a result, we get that a norm may apply to a given situation only to a certain extent and different norms may apply to different extents to the same situation.

4.1.1 Fuzzy Logic

Fuzzy logic was initiated by Lotfi Zadeh [65] with his seminal work on fuzzy sets. Fuzzy set theory provides a mathematical framework for representing and treating vagueness, imprecision, lack of information, and partial truth. Fuzzy logic is based on the notion of fuzzy set, a generalization of classical sets obtained by replacing the characteristic function of a set A, χ_A which takes up values in $\{0, 1\}$, i.e. $\chi_A(x) = 1$ iff $x \in A$, $\chi_A(x) = 0$ otherwise, with a *membership function* μ_A, which can take up any value in $[0, 1]$. The value $\mu_A(x)$ is the membership degree of element x in A, i.e., the degree to which x belongs in A. A fuzzy set is completely defined by its membership function. In fact, we can say that a fuzzy set *is* its membership function.

Operation on Fuzzy Sets The usual set-theoretic operations of union, intersection, and complement can be defined as a generalization of their counterparts on classical sets by introducing two families of operators, called triangular norms and triangular co-norms [52, 53, 42]. A triangular norm (or t-norm) is a binary operation $T : [0, 1] \times [0, 1] \to [0, 1]$ satisfying the following conditions for $x, y, z \in [0, 1]$:

- $T(x, y) = T(y, x)$ (commutativity);
- $T(x, T(y, z)) = T(T(x, y), z)$ (associativity):
- $y \leq z \Rightarrow T(x, y) \leq T(x, z)$ (monotonicity);
- $T(x, 1) = x$ (neutral element 1).

A well-known property about t-norms is:

$$T(x, y) \leq \min(x, y). \tag{1}$$

A triangular conorm (or t-conorm or s-norm), dual to a triangular norm, is a binary operation $S : [0, 1] \times [0, 1] \to [0, 1]$, whose neutral element is 0 instead of 1, with all other conditions identical to those of a t-norm:

- $S(x, y) = S(y, x)$ (commutativity);

- $S(x, S(y, z)) = S(S(x, y), z)$ (associativity):
- $y \leq z \Rightarrow S(x, y) \leq S(x, z)$ (monotonicity);
- $S(x, 0) = x$ (neutral element 0).

A well-known property about t-conorms is:

$$S(x, y) \geq \max(x, y). \tag{2}$$

If T is a t-norm, then $S(x, y) \equiv 1 - T(1 - x, 1 - y)$ is a t-conorm and *vice versa*: T and S in this case form a *dual pair* of a t-norm and a t-conorm. Noteworthy examples of such dual pairs are:

- $T_M(x, y) = \min\{x, y\}$, $S_M(x, y) = \max\{x, y\}$ (minimum t-norm and maximum t-conorm or Gödel t-norm and t-conorm);
- $T_P(x, y) = xy$, $S_P(x, y) = x + y - xy$ (product t-norm and t-conorm or probabilistic product and sum);
- $T_L(x, y) = \max\{x + y - 1, 0\}$, $S_L(x, y) = \min\{x + y, 1\}$ (Lukasiewicz t-norm and t-conorm or bounded sum);

For a given choice of a dual pair of a t-norm and a t-conorm (T, S), given two fuzzy sets A and B and an element x, the set-theoretic operations of union, intersection, and complement are thus defined as follows:

$$\mu_{A \cup B}(x) = S(\mu_A(x), \mu_B(x)); \tag{3}$$
$$\mu_{A \cap B}(x) = T(\mu_A(x), \mu_B(x)); \tag{4}$$
$$\mu_{\bar{A}}(x) = 1 - \mu_A(x). \tag{5}$$

4.2 Representing Norms

A norm r may be represented as a rule $b_1, \ldots, b_n \Rightarrow l$ such that l is the legal effect of r, such as an obligation linked to the norm [50]. A norm then has a conditional structure such as $b_1, \ldots, b_n \Rightarrow l$ (if $b_1, \ldots, b_n$ hold, then l ought to be the case). An agent is compliant with respect to this norm if l is obtained whenever $b_1, \ldots, b_n$ is derived. Often, logical models of legal reasoning assume that conditions of norms give a complete description of their applicability [50].

However, this assumption is too strong, due to the complexity and dynamics of the world. Norms cannot take into account all the possible conditions where they should or should not be applied, giving rise to the so called "penumbra": a core of cases which can clearly be classified as belonging to the concept. By a penumbra of hard cases, membership of the concept can be disputed. Moreover, not only does the world change as also pointed out in [36], giving rise to circumstances unexpected to the legislator who introduced the norm, but even the ontology of reality can change with respect to the one constructed by the law to describe the applicability conditions of norms. See, e.g., the problems concerning the application of existing laws to privacy, intellectual property or technological innovations in healthcare. To cope with unforeseen circumstances, the judicial system, at the moment in which a case concerning a violation is discussed in court, is empowered to interpret, i.e., to change norms, under some restrictions not to go beyond the purpose from which the norms stem.

The clauses of a norm often refer to imprecise concepts, which can take up different meanings depending on the purpose of the norm. The case for using fuzzy categories to account for such imprecise concepts has been made by da Costa Pereira *et al.* [17]: those imprecise concepts are a product of the human mind and, more precisely, of a categorization process. According to prototype theory, which is one of the most prominent and influential accounts of the cognitive processes of categorization, each category is defined by one or more prototypes [60], which are typical exemplars of it. A prototype may be regarded as being represented by a property list which has salient properties of the objects that are classified into the concept.

We may formalize these notions in a way that is compatible with an underlying knowledge representation standard and technical infrastructure like the ones provided by the W3C for the Semantic Web, i.e. OWL based on description logics for the terminological part and RDF for the assertional part. This would allow a practical implementation of our proposal using state-of-the-art knowledge engineering technologies. Nevertheless, we keep our formalization abstract for the sake of clarity.

Definition 11 (Language). *Given a knowledge base* K, *an* atom *is a unary or binary predicate of the form* $C(s)$, $R(s_1, s_2)$, *where the predicate symbol* C *is a concept name in* K *and* R *is a role name in* K, s, s_1, s_2 *are terms. A term is either a variable (denoted by* x, y, z) *or a constant (denoted by* a, b, c) *standing for an individual name or data value.*

According to this formalisation, an individual object o is described by all the facts of the form $C(o)$, $R(o, y)$ and $R(y, o)$ such that $K \models C(o)$, $K \models R(o, y)$ and $K \models R(y, o)$, where $\models$ stands for entailment. We call these facts the *properties* of o.

Definition 12 (Graded Category). *A graded category* $\tilde{C}$ *is described by a non-empty set of prototypes* $\mathrm{Prot}(\tilde{C}) = \{o_1, o_2, \ldots, o_n\}$, *where each* $o_i \in \mathrm{Prot}(\tilde{C})$ *is an individual name in* K.

We can consider that the choice of the actual (more plausible) category with respect to a prototype may be seen as if the prototype represented a kind of generalisation, which applied deductively, will allow to "classify" (categorise) new "problems" (instances) [7].

The membership of an instance to a category depends on its similarity to its prototype(s). Using a similarity measure with values in $[0,1]$ allows us to represent graded categories as fuzzy sets. A similarity measure of that kind may be defined. Here, we adapt the contrast model of similarity proposed by Tversky [58]. In such a model, an object is represented by means of a set of features and the similarity between two objects is defined as an increasing function of the features in common to the two objects, *common features*, and as a decreasing function of the features that are present in one object but not in the other, *distinctive features*.

Definition 13 (Number of Common Features). *Given two objects or individuals a, b in K, the number of their common features $c(a,b)$ is defined as*

$$\begin{aligned}
c(a,b) &= \|\{C : K \vdash C(a) \wedge C(b)\}\| \\
&+ \|\{\langle R, c \rangle : K \models R(a,c) \wedge R(b,c)\}\| \\
&+ \|\{\langle c, R \rangle : K \models R(c,a) \wedge R(c,b)\}\|,
\end{aligned}$$

where $\wedge$ represents the and *logical connective*.

Definition 14 (Number of Distinctive Features). *Given two objects or individuals a, b in K, the number of their distinctive features $dis(a,b)$ is defined as*

$$\begin{aligned}
dis(a,b) &= \|\{C : K \models C(a) \oplus C(b)\}\| \\
&+ \|\{\langle R, c \rangle : K \models R(a,c) \oplus R(b,c)\}\| \\
&+ \|\{\langle c, R \rangle : K \models R(c,a) \oplus R(c,b)\}\|,
\end{aligned}$$

where $\oplus$ represents the exclusive or *logical connective*.

It might be the case, in a given application, that some features are more important than others. This might be taken into account by defining different weights for each feature, depending on the application. Let $w : \text{Predicates} \to \mathbb{R}^+$ be a function associating a weight to each concept and role name in the language. The two functions c and dis might then be redefined as follows:

$$\begin{aligned}
c(a,b) &= \sum_{C:K \models C(a) \wedge C(b)} w(C) \\
&+ \sum_R w(R) \cdot \|\{c : K \models R(a,c) \wedge R(b,c)\}\| \\
&+ \sum_R w(R) \cdot \|\{c : K \models R(c,a) \wedge R(c,b)\}\|; \\
dis(a,b) &= \sum_{C:K \models C(a) \oplus C(b)} w(C) \\
&+ \sum_R w(R) \cdot \|\{c : K \models R(a,c) \oplus R(b,c)\}\| \\
&+ \sum_R w(R) \cdot \|\{c, : K \models R(c,a) \oplus R(c,b)\}\|.
\end{aligned}$$

These boil down to Definitions 13 and 14 when $w(C) = 1$ for all C and $w(R) = 1$ for all R.

Definition 15 (Object Similarity). *Given two objects or individuals a, b in K, their similarity is defined as*

$$s(a,b) = \frac{c(a,b)}{c(a,b) + \mathrm{dis}(a,b)}.$$

This similarity function satisfies a number of desirable properties. For all individuals a, b,

- $0 \leq s(a,b) \leq 1$;
- $s(a,b) = 1$ if and only if $a = b$;
- $s(a,b) = s(b,a)$;

We may now define the notion of membership degree of an object o in a graded category.

Definition 16. *Given a graded category $\tilde{C}$ and an arbitrary individual name o, the degree of membership of o in $\tilde{C}$ is given by*

$$\mu_{\tilde{C}}(o) = \underset{p \in \mathrm{Prot}(\tilde{C})}{S} s(o,p).$$

Since the category of an item in the left-hand-side of a rule may be vague or imprecise, the degrees of truth of such an item with respect to the actual situation may be partial. This implies that a rule can be partially activated, i.e., the state of affairs to be reached thanks to the compliance to that rule can be uncertain.

Let us consider the following rule r: $b_1, \ldots, b_n \Rightarrow l$, where the clauses b_i have the form "o_i is $\tilde{C}_i$" and let $\tilde{C}_1, \ldots, \tilde{C}_n$ be the categories of $b_1, \ldots, b_n$, respectively. A clause b_i of a norm involving a graded category may thus be true only to a degree. The premise of the norm may be partially true and a norm may thus apply only to some extent.

If the membership of an instance in a category depends on its similarity to the prototype of the category and also on the purpose of the norm, then we must conclude that both the prototype of a category and the similarity measure used to compute the membership might vary as a function of the purpose. While it may be hard to see how the similarity measure could change as a function of purpose, it is reasonable to assume that the legislators may have different prototypes in mind for a category with the same name when they write norms for different purposes.

This amounts to assuming that, given a graded category $\tilde{C}$, its set of prototypes may vary as a function of the purpose or goal G of the norm. We write $\mathrm{Prot}(\tilde{C} \mid G)$ to denote the set of the prototypes of category $\tilde{C}$ when the purpose of a norm is G.

The degree of truth α_{iG} of clause $b_i =$ "o_i is $\tilde{C}_i$", given that the purpose of the norm is G, may be computed as

$$\alpha_{iG} = \mu_{\tilde{C}_i}(o_i \mid G) = \underset{p \in \mathrm{Prot}(\tilde{C}|G)}{S}\, s(o_i, p). \qquad (6)$$

Definition 17. *The degree to which the premise $b_1, \ldots, b_n$ of rule of the form $b_1, \ldots, b_n \Rightarrow l$ is satisfied, given that the purpose of r is G, is given by*

$$\mathrm{Deg}(b_1, \ldots, b_n \Rightarrow l \mid G) = \underset{i=1,\ldots,n}{T}\, \alpha_{iG}.$$

The state of affairs which is reached thanks to the compliance of r will be associated with the truth degree of $\mathrm{Deg}(r \mid G)$ — this is also the degree associated to l after the activation of r.

5 Fuzzy Argumentation and Fuzzy Labeling

In recent years, several research efforts have attempted to combine formal argumentation and fuzzy logic, in such a way that the uncertainty of arguments can be measured by their fuzzy degrees, while the conflicts between arguments can be properly handled by Dung's argumentation semantics. Among them, Tamani and Croitoru [56] proposed a quantitative preference based argumentation system, called F-ASPIC. Based on ASPIC and fuzzy set theory, it can be used to model structured argumentation with fuzzy concepts. However, it is not clear how the status of a fuzzy argument is evaluated. Meanwhile, da Costa Perira *et al.* [18] introduce a labeling-based approach to evaluate the status of fuzzy arguments. Therefore, these two approaches are combined to lay a foundation for legal interpretation.

5.1 Fuzzy Argumentation System

A fuzzy argumentation system based on Tamani and Croitoru's F-ASPIC is proposed, with some adaptations to make it fit our framework, and with the addition of the fuzzy labeling algorithm proposed by [18].

The main differences between our framework and F-ASPIC [56] are as follows.

In our framework, we do not need to represent rules with different degrees of importance, as Tamani and Croitoru do. Unlike in F-ASPIC, the antecedent of a rule may be partially satisfied, if it involves graded categories. As a consequence, the consequent of that rule will have a partial truth degree and an argument depending on that rule has a partial membership in the set $\mathcal{A}$ of "active" arguments in the senese of da Costa Pereira et al.. So, although from a semantical point of view these gradual notions of partial truth or satisfaction are quite different from Tamani and Croitoru's notion of *importance* and *strength*, they lead to a mathematical treatment which is formally identical. Our main adaptation of F-ASPIC is therefore to replace, in the wording and in the formalism, these notions.

For the sake of simplicity, we assume that every element of the language and every rule are fallible. Hence, we do not differentiate between strict rules and defeasible rules, as ASPIC+ does, but we assume that we only have defeasible rules. This assumption makes the rationality postulates [2] trivially satisfied. However, it does not make things technically simpler (partial truth is basically preserved via strict rules, since they encode indisputable inferences). As a matter of fact, since strict rules satisfy contraposition (i.e., $P \Rightarrow Q$ is equivalent to $\neg Q \Rightarrow \neg P$), while defeasible rules do not have to, such behavior, when required, has to be explicitly simulated.

Definition 18 (Fuzzy argumentation system). *A fuzzy argumentation system, denoted as FAS, is a tuple $(\mathcal{L}, cf, \mathcal{R}, n, \mathrm{Deg})$ where*

- $\mathcal{L}$ *is a logical language.*

- *cf is a contrariness function (in this chapter, we only consider the classical negation $\neg$),*

- $\mathcal{R}$ *is the set of (defeasible) inference rules of the form $\phi_1, \ldots, \phi_m \Rightarrow \phi$ (where $\phi_i, \phi \in \mathcal{L}$) .*

- $n : \mathcal{R} \mapsto \mathcal{L}$ *is a naming convention for rules.*

- $\mathrm{Deg} : \mathcal{R} \to [0,1]$ *is a function returning the degree of activation of a rule, given a grounding of the formulas occurring in it. Intuitively, $\mathrm{Deg}(r)$ represents the degree of truth of the antecedent of r.*

In the original F-ASPIC system, fuzzy arguments are then constructed with respect to a fuzzy knowledge base $\mathcal{K}$, assigning a degree of importance $\mu_{\mathcal{K}}(p)$ to each proposition $p \in \mathcal{L}$. In our framework, however, we do not attach a degree of importance to propositions of formulas *per se*, but we need to evaluate a degree of truth of their grounding with respect to graded categories. To be more precise, the atomic propositions that are liable to have a partial degree of truth are those of the form "x is C", where C is a graded category. Given a substitution of variable x with an individual object o, the truth value of the grounding "o is C" will be given, as suggested in the previous section, by the similarity measure $s(o,p)$ of o to one of the prototypes p of C (i.e., one p in the set $\text{Prot}(C)$). To this aim, we keep the same symbol $\mathcal{K}$, but we regard it as a fuzzy valuation function.

Definition 19 (Fuzzy Valuation Function). *A fuzzy valuation function in a $FAS = (\mathcal{L}, cf, \mathcal{R}, n, \text{Deg})$ is a fuzzy set $\mathcal{K} : \mathcal{L}_{\text{ground}} \to [0,1]$ such that:*

- *if $\phi \in \mathcal{L}_{\text{ground}}$ is a ground atomic proposition of the form "o is C", with C a graded category,*

$$\mathcal{K}(o \text{ is } C) = \underset{p \in \text{Prot}(C)}{S}\, s(o,p); \tag{7}$$

- *if $\phi \in \mathcal{L}_{\text{ground}}$ is a ground atomic proposition not involving graded categories, $\mathcal{K}(\phi) \in \{0,1\}$;*

- *if $\phi, \psi \in \mathcal{L}_{\text{ground}}$,*

$$\begin{aligned}\mathcal{K}(\neg \phi) &= 1 - \mathcal{K}(\phi),\\ \mathcal{K}(\phi \wedge \psi) &= T(\mathcal{K}(\phi), \mathcal{K}(\psi))\\ \mathcal{K}(\phi \vee \psi) &= S(\mathcal{K}(\phi), \mathcal{K}(\psi))\end{aligned}$$

where T represents a triangular norm and S an associated triangular co-norm.

Let $r : b_1, \ldots, b_n \Rightarrow l$ be a rule. In a very simple case, the degree of activation Deg of r simply corresponds to the value returned by the Fuzzy Valuation Function $\mathcal{K}(\bigwedge_{1 \le k \le n} b_k)$.

Definition 20 (Fuzzy argument). *A fuzzy argument A on the basis of an argumentation theory with fuzzy valuation function $\mathcal{K}$ and a fuzzy argumentation system is*

- *ϕ if $\phi \in \mathcal{L}$ with: $Prem(A) = \{\phi\}$, $Conc(A) = \phi$, $Sub(A) = \{A\}$, $Rules(A) = \emptyset$.*

- *$A_1, \ldots, A_m \Rightarrow \phi$ if $A_1, \ldots, A_m$ are arguments such that there exists a rule $Conc(A_1), \ldots, Conc(A_m) \Rightarrow \psi$ in $\mathcal{R}$. In this case, $Prem(A) = Prem(A_1) \cup \cdots \cup Prem(A_m)$, $Conc(A) = \psi$, $Sub(A) = Sub(A_1) \cup \cdots \cup Sub(A_m) \cup \{A\}$, $Rules(A) = Rules(A_1) \cup \cdots \cup Rules(A_m) \cup \{Conc(A_1), \ldots, Conc(A_m) \Rightarrow \psi\}$.*

Given an argument A, $Conc(A)$ denotes the conclusion of A, $Prem(A)$ the set of the premises of A, $Sub(A)$ the set of the sub-arguments of A (including A itself), and $Rules(A)$ the set of rules involved in A.

Then, the degree of activation of each argument is measured by a fuzzy degree, called *strength of argument* in F-ASPIC, which can also be interpreted as a degree of membership in the set of active arguments, defined as follows.

Definition 21 (Strength of argument). *Given a fuzzy argument A, its strength, denoted $\mathcal{A}(A)$, is defined as follows:*

- *if A is of the form ϕ, then $\mathcal{A}(A) = \mathcal{K}(\phi)$;*

- *otherwise,*

$$\mathcal{A}(A) = \underset{r \in Rules(A)}{S} T\left(\mathrm{Deg}(r), \underset{\phi \in Prem(A)}{T} \mathcal{K}(\phi)\right). \qquad (8)$$

Then, with respect to the notions of rebut, undercut and defeat in ASPIC, the counterparts in the setting of fuzzy argumentation are defined as follows.

Unlike F-ASPIC, our framework does not require the definition of a fuzzy counterpart of the rebut, undercut, and defeat relation. We rely on the usual crisp relations, defined as follows.

Definition 22 (Attacks). *A attacks B iff A undercuts, rebuts or undermines B, where the function n is a naming convention for rules, which maps each rule to a well-formed formula in $\mathcal{L}$ [41], and*

- *A undercuts B (on B') iff $Conc(A) = \neg n(r)$ for some $B' \in Sub(B)$.*

- *A rebuts B (on B') iff $Conc(A) = \neg \phi$ for some $\exists B' \in Sub(B)$ of the form $B''_1, \ldots, B''_m \Rightarrow \phi$.*

- *A undermines B (on B') iff $Conc(A) = \neg \phi$ for some $B' = \phi$, $\phi \in Prem(B)$.*

Definition 23 (Defeat). *A defeats B iff A undercuts B on B', or A rebuts (undermines) B on B' and $\mathcal{A}(A) \not< \mathcal{A}(B')$.*

We use $\mathcal{A}$ and $\mathcal{D}$ to denote, respectively, the fuzzy set of active arguments (whose membership is their strength) and the defeat relation between them. Then, a fuzzy argumentation framework is represented as $\mathcal{F} = (\mathcal{A}, \mathcal{D})$.

This fuzzification of $\mathcal{A}$ provides a natural way of associating strengths to arguments, and suggests rethinking the labeling of an argumentation framework in terms of fuzzy degrees of argument acceptability [18]. The status of arguments can thus be evaluated by means of Fuzzy AF-labeling.

Definition 24 (Fuzzy AF-labeling). *Let $(\mathcal{A}, \mathcal{D})$ be a fuzzy argumentation framework. A fuzzy AF-labeling is a total function $\alpha \colon \mathcal{A} \mapsto [0, 1]$.*

Definition 25 (Fuzzy Reinstatement labeling). *Let $(\mathcal{A}, \mathcal{D})$ be a fuzzy argumentation framework, and α be a fuzzy AF-labeling. We say that α is a fuzzy reinstatement labeling iff, for all argument A,*

$$\alpha(A) = \min\{\mathcal{A}(A), 1 - \max_{B:(B,A)\in\mathcal{D}}\alpha(B)\} \tag{9}$$

Da Costa Periera et al. [18] made clear that given a fuzzy argumentation framework, its fuzzy reinstatement labeling may be computed by solving a system of n non-linear equations, where $n = \|\text{supp}(\mathcal{A})\|$, i.e., the number of arguments belonging to some non-zero degree in the fuzzy argumentation framework, of the same form as Equation 9, in n unknown variables, namely, the labels $\alpha(A)$ for all $A \in \text{supp}(\mathcal{A})$.

This can be done quite efficiently using an iterative method as follows: we start with an all-in labeling (a labeling in which every argument is labeled with the degree it belongs to $\mathcal{A}$). We denote by $\alpha_0 = \mathcal{A}$ this initial labeling, and by α_t the labeling obtained after the tth iteration of the labeling algorithm.

Definition 26. *Let α_t be a fuzzy labeling. An iteration in α_t is carried out by computing a new labeling α_{t+1} for all arguments A as follows:*

$$\alpha_{t+1}(A) = \frac{1}{2}\alpha_t(A) + \frac{1}{2}\min\{\mathcal{A}(A), 1 - \max_{B:(B,A)\in\mathcal{D}}\alpha_t(B)\}. \tag{10}$$

Note that Equation 10 guarantees that $\alpha_t(A) \leq \mathcal{A}(A)$ for all arguments A and for each step of the algorithm.

The above definition actually defines a sequence $\{\alpha_t\}_{t=0,1,...}$ of labelings, whose convergence has been proven [18]. We may now define the fuzzy labeling of a fuzzy argumentation framework as the limit of $\{\alpha_t\}_{t=0,1,...}$.

Definition 27. *Let $\langle \mathcal{A}, \mathcal{D} \rangle$ be a fuzzy argumentation framework. A fuzzy reinstatement labeling for such argumentation framework is, for all arguments A,*

$$\alpha(A) = \lim_{t \to \infty} \alpha_t(A). \tag{11}$$

Once this fuzzy reinstatement labeling has been computed, $\alpha(A)$ gives the degree to which each argument A in the framework is accepted; this degree may be used to compute the corresponding degree to which the purpose of a norm is G:

$$\alpha(G) = \max_{A:Conc(A)=G} \alpha(A). \tag{12}$$

As it is clear from the above definitions, an argument may be accepted partially and thus the purpose of a norm may be uncertain. Now, different strategies may be used to deal with such an uncertainty. One possibility is to consider the purpose G for which $\alpha(G)$ is maximal. Another is to evaluate the norm with respect to all purposes such that $\alpha(G) > 0$ and then combine the results weighted by ther corresponding $\alpha(G)$.

6 Interpreting a Norm with Flexibility

In addition to taking graded categories into account, any norm is always associated with a purpose: that is what is called the *purpose* of the norm. The idea is then to capture the fact that, when a legislator states a norm, she has in mind a state of affairs to be reached through compliance with that norm. With that in mind, the degree to which a concept in the rule belongs to a category would also depend on the purpose associated with the rule. In other words, given a norm like $b_1, \ldots, b_n \Rightarrow l$, the degree associated to l depends on the degrees of truth of conditions b_i. These degrees depend in turn on the purpose associated to the norm: for example, the greater the extent to which the prohibition to smoke in public spaces promotes the goal *public health*, the greater is the degree of applicability of a rule like Public_Space $\Rightarrow$ No_Smoking assuming the fuzziness of the concept Public_Space. However, the actual purpose of the legislator can be controversial [36]: for example, not enough evidence or factual information might be available which could help discover what the legislator was intending when writing a norm. Note that the historical purpose could be obsolete due to social, economic or political change, and the legislator has not reacted in a timely manner or at all. Here, as done in legal theory [43, 50], we adopt an objective teleological approach to interpretation, which means that the purpose of a norm is the one that any rational interpreter would assign to it. Hence, we use an argumentative system which will determine which purpose, with respect to the current knowledge, is the most plausible purpose of a norm.

The case study in our chapter is the application of the Italian Legislative Act n. 40/2004 on "Medically Assisted Reproduction." Before the declaration of uncostitutionality ruled by the Constitutional Court (opinion n. 96/2015), the statute included section 4, par. 1: "The recourse to medically assisted reproduction techniques is allowed only [...] in the cases of sterility or infertility [...]." The purpose of the discussion is to see whether this provision can be interpreted so that non-sterile or fertile couples, in which one or both spouses are immune carriers of a serious genetic anomaly, could access those techniques.

These couples are able to conceive and bear a child, though the probability that the baby will contract the disease is high. These diseases are normally severely disabling, provoke physical dysfunctions, often prevent the full psychological development of the baby, and can cause premature death. The mentioned medical techniques can detect the illness in advance and consequently let the parents take aware decisions about the pregnancy.

The legislative act does not explicitly define 'sterility' and 'infertility.' On the basis of art. 7 l. 40/2004, every three years, the Ministry of Health is required to promulgate a decree containing the updated guidelines for the application of the law. According to these guidelines, the terms 'sterility' and 'infertility' are considered synonyms and refer to the lack of conception, in addition to those cases of certified pathology, after 12/24 months of regular sexual relations in a heterosexual couple.

In civil law systems, when it comes to statutory interpretation, one option is teleological interpretation, according to which, when interpreting a provision, judges often take into account what explicit or implicit purposes can be ascribed to the norm [43, 36].

As for the purposes, law n. 40/2004 states as follows:

Art. 1, on "Purposes". Par.1: In order to favour the solution of reproduction problems caused by human sterility or infertility, it is allowed the recourse to medically assisted reproduction techniques, according to the conditions and the modalities provided for by the present law, which guarantees the respect of the rights of all the subjects involved, included the conceived baby.

Let us also consider the following norm from art. 4 of L. n. 40/2004:

> The use of techniques of medically assisted procreation is [...] confined to the cases with issue of infertility or [...] sterility certified by a medical procedure.

Law n. 40/2004 is connected to other statutes of the legal system. In particular, the Italian Legislative Act n.194/1978 on "Social Protection of Maternity and Abortion" provides for the possibility of a therapeutic abortion if, during pregnancy, a pathological condition is ascertained, including those relating to significant anomalies or malformations of the baby, that put at risk the physical or psychic health of the woman." Severe genetic diseases are thus included. Moreover, along law n. 194/1978, the chance of a serious danger for the life of the woman is seen as a reason to proceed to abortion. This second legislative act is thus meant to promote the right to health both of the mother and of the child.

In light of the previous remarks, we can outline a list of interpretive arguments supporting different interpretations. Our main target is to see what interpretation better promotes the purposes that can be ascribed to the norm, if a purpose can be considered prominent, and what attacks can occur.

In what follows we present a plausible set of rules representing norms and interpretive legal arguments about such norms [49]. In both cases, fuzzy argumentation is related to the promotion of legal purposes.

In particular, the following (defeasible) rules can identify the basic the interpretive arguments arg_1, arg_2, arg_3, respectively, at stake:

$$r_1: \neg\text{Ste}(x), \text{Rsn\_Exp\_Life}(x) \Rightarrow \neg\text{Med\_Rpr}(x)$$
$$r_2: \text{Med\_Rpr}(x), \text{Genetic\_Dis}(x), \text{Well\_Being}(x) \Rightarrow \text{Sol\_Rep\_Prob}(x)$$
$$r_3: \neg\text{Sol\_Rep\_Prob}(x), \text{Genetic\_Dis}(x) \Rightarrow \neg\text{Rsn\_Exp\_Life}(x)$$
$$r_4: \text{Gener\_Child}(x) \Rightarrow \neg\text{Ste}(x)$$

where

- $\text{Ste}(x)$ = "x is sterile",

- $\text{Med\_Rpr}(x)$ = "x can access to medically assisted reproduction techniques",

- $\text{Rsn\_Exp\_Life}(x)$ = "x grants a reasonably expected life",

- $\text{Genetic\_Dis}(x)$ = "x is affected by a serious genetic disease",

- $\text{Well\_Being}(x)$ = "x enjoys psychological well-being",

- $\text{Sol\_Rep\_Prob}(x)$ = "legally solved for x the reproduction problems",

- $\text{Gener\_Child}(x)$ = "x can generate children".

Consider the case mentioned above: a couple is actually able to conceive and generate children ($\text{Gener\_Child}(CP)$), but they are both carriers of a serious genetic disease ($\text{Genetic\_Dis}(CP)$), which does not allow children to live for more than a few years. Then according to the above rules, we have the following arguments:

$$arg_1 = \neg\text{Sol\_Rep\_Prob(CP)}, \text{Genetic\_Dis(CP)} \Rightarrow$$
$$\neg\text{Rsn\_Exp\_Life(CP)}$$
$$arg_2 = \text{Gener\_Child(CP)} \Rightarrow \neg\text{Ste(CP)} \Rightarrow$$
$$\text{Rsn\_Exp\_Life(CP)}, \neg\text{Ste(CP)} \Rightarrow \neg\text{Med\_Rpr(CP)}$$
$$arg_3 = \text{Med\_Rpr(CP)}, \text{Genetic\_Dis(CP)},$$
$$\text{Well\_Being(CP)} \Rightarrow \text{Sol\_Rep\_Prob(CP)}.$$

The attack relation between arguments are: arg_1 attacks arg_2, arg_2 attacks arg_3, and arg_3 attacks arg_1. Then, we have the following argumentation framework:

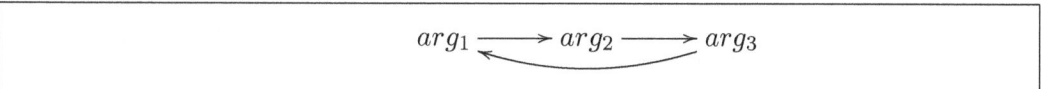

Figure 2: An argumentation framework

Let us consider these purposes:

- **Hlth_Of_MnC** ="purpose: the right to health both of the mother and the child"; this purpose is associated to rule r_2, i.e., we assume that r_2 promotes purpose **Hlth_Of_MnC**;

- **No_Eugenic** ="purpose: no eugenic selection"; this purpose is associated to rules r_1 and r_4, i.e., we assume that r_1 and r_4 promote purpose **No_Eugenic**.

For the sake of illustration, let us also assume that only two concepts are fuzzy: Gener_Child and Well_Being. Hence, if we consider, for example, r_4, this means that fuzziness depends only on the fact that rule r_4 makes the degree of ¬Ste(CP) as dependent on the degree of capability of generating children by CP. No other source of vagueness are considered for r_4. Analogous considerations apply to rule r_2 in regard to Well_Being.

Given these purposes, we can measure the *degrees to which the premise* of rules r_2 and r_4 are satisfied by CP.

- **Rule r_4:** Let us assume that only one prototype p_1 is associated to Gener_Child and **No_Eugenic** (for example, a standard fertile couple statistically identified in the population of couples) in which, among others, the expected life of children is greater than 50 years and the incidence of genetic diseases is less than 20%. Clearly, these are distinctive features that differentiates p_1 with respect to CP: suppose that the overall distinctive features are $d_1, \ldots, d_6$, while the common features are $c_1, \ldots, c_4$.

If we apply Definition 15, then $s(\text{CP}, p_1) = \frac{4}{4+6} = \frac{4}{10} = 0.4$. Since p is the unique prototype for Gener_Child with respect to **No_Eugenic** and that G for r_4 is $\{\textbf{No\_Eugenic}\}$, then it is easy to check that (see, in particular, Definitions 16 and 19)

$$\mu_{\text{Gener\_Child}}(\text{CP}) = \text{Deg}(r_4 \mid G) = \mathcal{K}(\text{Gener\_Child}(\text{CP})) = 0.4.$$

- **Rule r_2:** Let us assume that only one prototype p_2 is associated to Well_Being and **Hlth_Of_MnC** and that the overall distinctive features are $d'_1, \ldots, d'_1 6$, while the common features are $c'_1, \ldots, c'_4$. For the same reason, given that $A(r_2)$ stands for Med_Rpr(CP) $\wedge$ Genetic_Dis(CP) $\wedge$ Well_Being(CP),

$$s(\text{CP}, p_2) = \mu_{\text{Well\_Being}}(\text{CP}) = \text{Deg}(r_2 \mid G') = \mathcal{K}(A(r_2)) = 0.2.$$

Given these degrees of activation of rules, the following table illustrates how to apply the machinery of fuzzy labeling to this scenario, given the above degrees of activation of the rules that determine the strength of arguments. As we noted, we defined the fuzzy labeling of a fuzzy argumentation framework as the limit of $\{\alpha_t\}_{t=0,1,\ldots}$. The convergence is obtained quickly: a small number of iterations is enough to get close to the limit.

t	$\alpha_t(\textbf{arg1})$	$\alpha_t(\textbf{arg2})$	$\alpha_t(\textbf{arg3})$
0	1	0.4	0.2
1	0.9	0.2	0.2
2	0.85	0.15	0.2
3	0.825	0.15	0.2
4	0.8125	0.1625	**0.2**
5	**0.8**	0.175	↓
6	↓	**0.2**	

Table 1: Fuzzy labeling

Therefore, arg_1 is accepted to degree 0.8 while arg_2 and arg_3 are given a much lower acceptance degree, namely 0.2. In other words, arg_1 is much more acceptable than arg_2 and arg_3. Its important to observe that these degrees just represent an order of plausibility, as if saying that arg_1 is four times as plausible as arg_2 or arg_3.

7 Related work

Young et al. [64] endowed Brewka's prioritized default logic (PDL) with argumentation semantics using the $ASPIC^+$ framework for structured argumentation [41]. More precisely, their goal is to define a preference ordering over arguments $\succsim$, based on the strict total order over defeasible rules defined to instantiate $ASPIC^+$ to PDL, so as to ensure that an extension within PDL corresponds to the justified conclusions of its $ASPIC^+$ instantiation. Several options are investigated, and they demonstrate that the standard $ASPIC^+$ *elitist* ordering cannot be used to calculate $\succsim$ as there is no correspondence between the argumentation-defined inferences and PDL, and the same holds for a disjoint elitist preference ordering. The authors come up with a new argument preference ordering definition which captures both preferences over arguments and also *when* defeasible rules become applicable in the arguments' construction, leading to the definition of a strict total order on defeasible rules and corresponding non-strict arguments. Their representation theorem shows that a correspondence always exists between the inferences made in PDL and the conclusions of justified arguments in the $ASPIC^+$ instantiation under stable semantics.

Brewka and Eiter [15] consider programs supplied with priority information, which is given by a supplementary strict partial ordering of the rules. This additional information is used to solve potential conflicts. Moreover, their idea is that conclusions should be only those literals that are contained in at least one answer set. They propose to use preferences on rules for selecting a subset of the answer sets, called the *preferred answer sets*. In their approach, a rule is applied unless it is defeated via its assumptions by rules of higher priorities.

Dung [21] presents an approach to deal with contradictory conclusions in defeasible reasoning with priorities. More precisely, he starts from the observation that often, the proposed approaches to defeasible reasoning with priorities (e.g., [14, 51, 40]) sanction contradictory conclusions, as exemplified by $ASPIC^+$ using the weakest link principle together with the elitist ordering which returns contradictory conclusions with respect to its other three attack relations, and the conclusions reached with the well known approach of Brewka and Eiter [15]. Dung shows then that the semantics for any complex interpretation of default preferences can be characterized by a subset of the set of stable extensions with respect to the normal attack relation assignments, i.e., a normal form for ordinary attack relation assignments. Dung's *normal attack relation* satisfies some desirable properties (Credulous cumulativity and Attack monotonicity) that cannot be satisfied by the $ASPIC^+$ semantics [21], i.e., the semantics of structured argumentation with respect to a given ordering of structured arguments (elitist or democratic pre-order) in $ASPIC^+$. In the setting of this paper,

3065

this notion could be defined as follows. Let $\alpha = (a_1, \ldots, a_n)$ and $\beta = (b_1, \ldots, b_m)$ be arguments constructed from a hierarchical abstract normative system. Since we have no Pollock style undercutting argument (as in $ASPIC^+$) and each norm is assumed to be defeasible, α is said to normally attack argument β if and only if β has a sub-argument β' such that $concl(\alpha) = \overline{concl(\beta')}$, and $r((a_{n-1}, a_n)) \geq r((b_{m-1}, b_m))$. According to the weakest link principle and Definition 23, the normal defeat relation is equivalent to the defeat relation using the last link principle in this paper.

Kakas et al. [32] present a logic of arguments called *argumentation logic*, where the foundations of classical logical reasoning are represented from an argumentation perspective. More precisely, their goal is to integrate into the single argumentative representation framework both classical reasoning, as in propositional logic, and defeasible reasoning.

You et al. [63] define a prioritized argumentative characterization of nonmonotonic reasoning, by casting default reasoning as a form of prioritized argumentation. They illustrate how the parameterized formulation of priority may be used to allow various extensions and modifications to default reasoning.

We, and all these approaches, share the idea that an argumentative characterization of NMR formalisms, like prioritized default logic in Young's case and hierarchical abstract normative systems in our approach, contributes to make the inference process more transparent to humans. However, the targeted NMR formalism is different, leading to different challenges in the representation results. To the best of our knowledge, no other approach addressed the challenge of an argumentative characterization of prioritized normative reasoning.

Prakken and Sartor [48] proposed to define a dynamic argumentation system as a tuple $S = \langle \mathcal{L}, -, \mathcal{R}, n \rangle$ where $\mathcal{L}$ is a logical language including symbols for predicates, functions, constants and variables, = for equality, ¬ for negation and $\rightsquigarrow$ for normative conditionals, and the universal quantifier $\forall$, $\mathcal{R}$ is the set of inference rules, and n is the naming convention. A norm has the form $\forall (L_1 \wedge \ldots \wedge L_n \rightsquigarrow L)$, where $L_1, \ldots, L_n$ are literals. In particular, they define inference schemes for validity $(Valid(N(\phi)) \rightarrow \phi)$, and applicability (i.e., undercutting, $\neg Applicable(w) \rightarrow \neg DMP(w)$). As future direction, the authors foster the extension of the framework by enriching the logical language with a formal account of modalities such as obligation. This is the issue we addressed in this chapter.

Van der Torre and Villata [59] extend their dynamic legal argumentation framework with deontic modalities, and they propose an general framework for legal reasoning based on ASPIC-like argumentation and input/output logic. The framework allows to reason over normative concepts like factual and deontic detachment, and to assess norms' equivalence. The properties of our logical framework are proved. All new concepts are illustrated by a running example. Our main technical contribution is to give a formal analysis of legal argumentation, and a bridge to standard formalisms for normative systems like input/output logic. Compared to other input/output logics, van der Torre and Villata do not have weakening of the output or aggregation of obligations due to the clausal language. For a comparison with other deontic logics in the recent handbook on deontic logic and normative systems we can define the inference relation in terms of consequence sets as usual (e.g., $KB \models \phi$ iff $\phi \in Out(KB)$).

A framework for legal interpretation capable of taking graded, purpose-dependent institutional facts into account has been proposed by da Costa Periera et al. [17]. Such a framework uses argumentation to handle conflicts between different interpretations of legal concepts. The originality of this proposal lies in the use of argumentation to identify the most likely purpose of a norm, which in turn circumscribes the interpretation of the categories (institutional facts, legal concepts) referred to by the norm. The idea of using many-valued logics in argumentation theory is not new. Just to name a few, [16] define a notion of gradual acceptability such that a numerical value is assigned to each argument on the basis of its attackers; Janssen et al. [31] propose a fuzzy approach enriching the expressive power of classical argumentation, whose originality lies in the fact that the framework allows to represent the relative strength of the attacks; Grossi and Modgil [26] propose a graded generalization of argumentation semantics in which the origin of the justification degrees is supposed to be exclusively endogenous, i.e., based exclusively on the topology of the attack relation. Qualitative approaches to arguments' acceptability have been proposed in preference-based argumentation frameworks (*PAF*) [3], value-based argumentation frameworks (*VAF*) [9], and weighted argumentation frameworks (*WAF*) [22]. These approaches do not define graded semantics: *(i) PAF*s take into account preference orderings in the selection of acceptable conflicting arguments; *(ii) VAF*s are based on the assumption that some arguments can be stronger than others with respect to a certain value they advance, and this affects the success of an attack; and *(iii)* in *WAF*s, the weights are used for deciding which attacks can be ignored when computing the extensions. In these approaches, however, preference, values, and weights are provided only as input for the computation of extensions; they do not return an acceptability degree for arguments as output. Finally, Gabbay [23] proposes an equational approach which returns multiple (graded) solutions, and thus several rankings for one argumentation framework.

Other frameworks for legal argumentation are listed below, but all of them concentrate on specific problems of reasoning with legal arguments, whilst the aim of our framework, as well as of Prakken and Sartor, is to integrate various aspects so far addressed separately towards a logic comprehensive model of dynamic legal argumentation. The combination of inferences establishing the validity of norms with inferences using valid norms has been proposed by Yoshino [62]. The view that valid norms are defeasible reasons for legal conclusions was at the core of reason based logic by Hage [27]. Arguments about applicability and inapplicability of norms are discussed by Gordon, Prakken and colleagues [24, 47]. Modeling reasoning with norms through argumentation schemes has been formalized by Verheij [61]. Further connections between norms and argumentation include, among others, case based reasoning [6], arguing in rule based systems [45, 47], dialogues and dialectics [24], argument schemes [25, 11].

Several works in the literature of AI & Law have considered the role of purposes in the legal interpretation. Indeed, this idea is standard in legal theory and the purpose of legal rules is recognised by jurists as decisive in clarifying the scope of the legal concepts that qualify the applicability conditions for those rules [8, 46, 55, 27]. [8, 46] use purposes/goals and values in frameworks of case based reasoning for modeling precedents mainly in a common law context. [55] analyse a number of legal arguments even in statutory law, which include cases close to the ones discussed here. Hage [27] addresses, among others, the problem of reconstructing extensive and restrictive interpretation. This is done in Reason-Based Logic, a logical formalism that can deal with rules and reasons: the idea is that the satisfaction of rules' applicability conditions is usually a reason for application of these rules, but there can also be other (and possibly competing) reasons, among which we have the goals that led the legislator to make the rules. More recently, various work [12, 13, 66] proposed formal models for teleological interpretation in statutory law. All these approaches in AI & Law highlight the importance of rule purposes/goals. However, it seems that no work so far has attempted to couple this view with fuzzy logic and argumentation. In this perspective, we believe that this chapter may contribute to fill a gap in the literature.

8 Conclusions

In this article, we discuss three examples from the literature of handling norms by means of formal argumentation. First, we discuss how the so-called Greedy and Reduction approaches can be represented using the weakest and the last link principles respectively [35]. Based on such representation results, formal argumentation can be used to explain the detachment of obligations and permissions from hierarchical normative systems in a new way. Second, we discuss a dynamic ASPIC-based legal argumentation theory [48], and we discuss how existing logics of normative systems can be used to analyse such new argumentation systems [59]. Third, we show how argumentation can be used to reason about other challenges in normative systems as well, by discussing a model for arguing about legal interpretation [17]. In particular, we show how fuzzy logic combined with formal argumentation can be used to reason about the adoption of graded categories and thus address the problem of open texture in normative interpretation. We refer to the original papers for further details.

Our aim to discuss these three examples is to inspire new applications of formal argumentation to the challenges of normative reasoning in multiagent systems. We do not assume that the possible interactions between normative reasoning and formal argumentation is restricted to the three examples we discuss in this article. Besides resolving conflicting norms, norm compliance, norm dynamics and norm interpretation, it has been used also to argue about enforced obligations and permissions, and to establish norms' validity by deriving their conclusions. Moreover, other central challenge in normative multi-agent system are discussed in the article of Pigozzi and van der Torre, and we believe that formal argumentation is also applicable to various other challenges. For example, agents can argue about the creation or emerging of norms from the mental states of individual agents, or how normative systems can be merged.

References

[1] Carlos E. Alchourron and David Makinson. Hierarchies of regulations and their logic. In Risto Hilpinen, editor, *New studies in deontic logic*, pages 125–148. Springer, 1981.

[2] Leila Amgoud. Postulates for logic-based argumentation systems. *Int. J. Approx. Reasoning*, 55(9):2028–2048, 2014.

[3] Leila Amgoud and Claudette Cayrol. Integrating preference orderings into argument-based reasoning. In Dov M. Gabbay, Rudolf Kruse, Andreas Nonnengart, and Hans Jürgen Ohlbach, editors, *Qualitative and Quantitative Practical Reasoning, First International Joint Conference on Qualitative and Quantitative Practical Reasoning ECSQARU-FAPR'97, Bad Honnef, Germany, June 9-12, 1997, Proceedings*, volume 1244 of *Lecture Notes in Computer Science*, pages 159–170. Springer, 1997.

[4] Michal Araszkiewicz and Tomasz Zurek. Comprehensive framework embracing the complexity of statutory interpretation. In *Legal Knowledge and Information Systems - JURIX 2015: The Twenty-Eighth Annual Conference, Braga, Portual, December 10-11, 2015*, pages 145–148, 2015.

[5] Michal Araszkiewicz and Tomasz Zurek. Interpreting agents. In *Legal Knowledge and Information Systems - JURIX 2016: The Twenty-Ninth Annual Conference*, pages 13–22, 2016.

[6] Kevin D. Ashley. *Modeling legal argument - reasoning with cases and hypotheticals*. Artificial Intelligence and Legal Reasoning. MIT Press, 1990.

[7] Kevin D. Ashley. Reasoning with cases and hypotheticals in HYPO. *International Journal of Man-Machine Studies*, 34(6):753–796, 1991.

[8] Trevor J. M. Bench-Capon. The missing link revisited: The role of teleology in representing legal argument. *Artif. Intell. Law*, 10(1-3):79–94, 2002.

[9] Trevor J. M. Bench-Capon. Value-based argumentation frameworks. In Salem Benferhat and Enrico Giunchiglia, editors, *9th International Workshop on Non-Monotonic Reasoning (NMR 2002), April 19-21, Toulouse, France, Proceedings*, pages 443–454, 2002.

[10] Trevor J. M. Bench-Capon, Henry Prakken, and Giovanni Sartor. *Argumentation in Legal Reasoning*. Argumentation in Artificial Intelligence. Springer, 2010.

[11] Floris Bex, Henry Prakken, Chris Reed, and Douglas Walton. Towards a formal account of reasoning about evidence: Argumentation schemes and generalisations. *Artif. Intell. Law*, 11(2-3):125–165, 2003.

[12] Guido Boella, Gguido Governatori, Antonino Rotolo, and Leendert W. N. van der Torre. *Lex Minus Dixit Quam Voluit, Lex Magis Dixit Quam Voluit*: A formal study on legal compliance and interpretation. In P. Casanovas, U. Pagallo, G. Sartor, and G. Ajani, editors, *AICOL-I/IVR-XXIV and AICOL-II/JURIX 2009 Revised Selected Papers*, pages 162–183, 2009.

[13] Guido Boella, Guido Governatori, Antonino Rotolo, and Leendert W. N. van der Torre. A logical understanding of legal interpretation. In F. Lin, U. Sattler, and M. Truszczynski, editors, *Proceedings of the Twelfth International Conference on the Principles of Knowledge Representation and Reasoning (KR 2010)*, 2010.

[14] Gerhard Brewka. Preferred subtheories: An extended logical framework for default reasoning. In N. S. Sridharan, editor, *Proceedings of the 11th International Joint Conference on Artificial Intelligence. Detroit, MI, USA, August 1989*, pages 1043–1048. Morgan Kaufmann, 1989.

[15] Gerhard Brewka and Thomas Eiter. Preferred answer sets for extended logic programs. *Artificial Intelligence*, 109(1-2):297–356, June 1999.

[16] Claudette Cayrol and Marie-Christine Lagasquie-Schiex. Graduality in argumentation. *J. Artif. Intell. Res. (JAIR)*, 23:245–297, 2005.

[17] Célia da Costa Pereira, Andrea Tettamanzi, Beishui Liao, Alessandra Malerba, Antonino Rotolo, and Leendert van der Torre. Combining fuzzy logic and formal argumentation for legal interpretation. In Guido Governatori, editor, *Proceedings of the 16th International Conference on Artificial Intelligence and Law, ICAIL 2017, London, UK, June 12-16, 2017*. ACM, 2017.

[18] Célia da Costa Pereira, Andrea G. B. Tettamanzi, and Serena Villata. Changing one's mind: Erase or rewind? possibilistic belief revision with fuzzy argumentation based on trust. In Toby Walsh, editor, *Proceedings of the Twenty-Second International Joint Conference on Artificial Intelligence (IJCAI'11), Barcelona, Catalonia, Spain, July 16-22, 2011*, pages 164–171. AAAI, 2011.

[19] Anthony D'Amato. Legal uncertainty. *California Law Review*, 71(1):1–55, 1983.

[20] Phan Minh Dung. On the acceptability of arguments and its fundamental role in nonmonotonic reasoning, logic programming and n-person games. *Artif. Intell.*, 77(2):321–358, 1995.

[21] Phan Minh Dung. An axiomatic analysis of structured argumentation with priorities. *Artif. Intell.*, 231:107–150, 2016.

[22] Paul E. Dunne, Anthony Hunter, Peter McBurney, Simon Parsons, and Michael Wooldridge. Weighted argument systems: Basic definitions, algorithms, and complexity results. *Artif. Intell.*, 175(2):457–486, 2011.

[23] Dov M. Gabbay. Equational approach to argumentation networks. *Argument & Computation*, 3(2-3):87–142, 2012.

[24] Thomas F. Gordon. The pleadings game. *Artif. Intell. Law*, 2(4):239–292, 1993.

[25] Thomas F. Gordon and Douglas Walton. Legal reasoning with argumentation schemes. In *Proceedings of the Twelfth International Conference on Artificial Intelligence and Law (ICAIL 2009)*, pages 137–146, 2009.

[26] Davide Grossi and Sanjay Modgil. On the graded acceptability of arguments. In Qiang Yang and Michael Wooldridge, editors, *Proceedings of the Twenty-Fourth International Joint Conference on Artificial Intelligence, IJCAI 2015, Buenos Aires, Argentina, July 25-31, 2015*, pages 868–874. AAAI Press, 2015.

[27] Jaap Hage. *Reasoning with Rules: An Essay on Legal Reasoning and Its Underlying Logic*. Kluwer, 1997.

[28] Jörg Hansen. Prioritized conditional imperatives: problems and a new proposal. *Autonomous Agents and Multi-Agent Systems*, 17(1):11–35, 2008.

[29] Herbert L. A. Hart. *The Concept of Law*. Clarendon Press, Oxford, 1994.

[30] Philipp Heck. *Begriffsbildung und Interessensjurisprudenz*. Mohr Siebeck, Tübingen, 1932.

[31] Jeroen Janssen, Martine De Cock, and Dirk Vermeir. Fuzzy argumentation frameworks. In *Procedings of the 12th International Conference on Information Processing and Management of Uncertainty in Knowledge-Based Systems (IPMU 2008)*, pages 513–520, 2008.

[32] Antonis C. Kakas, Francesca Toni, and Paolo Mancarella. Argumentation for propositional logic and nonmonotonic reasoning. In Laura Giordano, Valentina Gliozzi, and Gian Luca Pozzato, editors, *Proceedings of the 29th Italian Conference on Computational Logic, Torino, Italy, June 16-18, 2014.*, volume 1195 of *CEUR Workshop Proceedings*, pages 272–286. CEUR-WS.org, 2014.

[33] George Lakoff. *Women, Fire, and Dangerous Things*. University of Chicago Press, Chicago, 1987.

[34] George Lakoff and Mark Jonhson. *Metaphors We Live By*. University of Chicago Press, Chicago, 1980.

[35] Beishui Liao, Nir Oren, Leendert van der Torre, and Serena Villata. Prioritized norms and defaults in formal argumentation. In *Proceedings of the 13th International Conference on Deontic Logic and Normative Systems (DEON2016)*, pages 139–154, 2016.

[36] Doris Liebwald. Law's capacity for vagueness. *Int J Semiot Law*, 26:391–423, 2013.

[37] D.N. MacCormick and R.S. Summers, editors. *Interpreting Statutes: A Comparative Study*. Ashgate, 1991.

[38] David Makinson and Leendert van der Torre. Input/output logics. *J. Philosophical Logic*, 29(4):383–408, 2000.

[39] Alessandra Malerba, Antonino Rotolo, and Guido Governatori. Interpretation across legal systems. In *Legal Knowledge and Information Systems - JURIX 2016: The Twenty-Ninth Annual Conference*, pages 83–92, 2016.

[40] Sanjay Modgil and Henry Prakken. A general account of argumentation with preferences. *Artif. Intell.*, 195:361–397, 2013.

[41] Sanjay Modgil and Henry Prakken. The $ASPIC^+$ framework for structured argumentation: a tutorial. *Argument & Computation*, 5(1):31–62, 2014.

[42] Mirko Navara. Triangular norms and conorms. *Scholarpedia*, 2(3):2398, 2007. revision #137537.

[43] Aleksander Peczenik. *On Law and Reason*. Kluwer, 1989.

[44] Gabriella Pigozzi and Leendert van der Torre. Arguing about constitutive and regulative norms. *Journal of applied nonclassical logics*, to appear.

[45] Henry Prakken. A logical framework for modelling legal argument. In *Proceedings of the Fourth International Conference on Artificial intelligence and Law (ICAIL 1993)*, pages 1–9, 1993.

[46] Henry Prakken. An exercise in formalising teleological case-based reasoning. *Artif. Intell. Law*, 10:113–133, 2002.

[47] Henry Prakken and Giovanni Sartor. A dialectical model of assessing conflicting arguments in legal reasoning. *Artif. Intell. Law*, 4(3-4):331–368, 1996.

[48] Henry Prakken and Giovanni Sartor. Formalising arguments about norms. In Kevin D. Ashley, editor, *JURIX*, volume 259 of *Frontiers in Artificial Intelligence and Applications*, pages 121–130. IOS Press, 2013.

[49] Antonino Rotolo, Guido Governatori, and Giovanni Sartor. Deontic defeasible reasoning in legal interpretation: two options for modelling interpretive arguments. In Ted Sichelman and Katie Atkinson, editors, *ICAIL 2015*, pages 99–108, 2015.

[50] Giovanni Sartor. *Legal reasoning: A cognitive approach to the law*, volume 5 of *A Treatise of Legal Philosophy and General Jurisprudence*. Springer, Berlin, 2005.

[51] Torsten Schaub and Kewen Wang. A comparative study of logic programs with preference. In Bernhard Nebel, editor, *Proceedings of the Seventeenth International Joint Conference on Artificial Intelligence, IJCAI 2001, Seattle, Washington, USA, August 4-10, 2001*, pages 597–602. Morgan Kaufmann, 2001.

[52] Berthold Schweizer and Abe Sklar. Statistical metric spaces. *Pacific J. Math.*, 10(1):313–334, 1960.

[53] Berthold Schweizer and Abe Sklar. *Probabilistic metric spaces*. North Holland series in probability and applied mathematics. North Holland, 1983.

[54] Scott J. Shapiro. *Legality*. Harvard University Press, 2011.

[55] David B. Skalak and Edwina L. Rissland. Arguments and cases: An inevitable intertwining. *Artif. Intell. Law*, 1:3–44, 1992.

[56] Nouredine Tamani and Madalina Croitoru. A quantitative preference-based structured argumentation system for decision support. In *Proceedings of the 2014 IEEE International Conference on Fuzzy Systems (FUZZ-IEEE)*, pages 1408–1415, 2014.

[57] Silvano Colombo Tosatto, Pierre Kelsen, Qin Ma, Marwane El Kharbili, Guido Governatori, and Leendert W. N. van der Torre. Algorithms for tractable compliance problems. *Frontiers of Computer Science*, 9(1):55–74, 2015.

[58] Amos Tversky. Features of similarity. *Psychological Review*, 84(4):327–352, 1977.

[59] Leendert W. N. van der Torre and Serena Villata. An aspic-based legal argumentation framework for deontic reasoning. In Simon Parsons, Nir Oren, Chris Reed, and Federico Cerutti, editors, *Computational Models of Argument - Proceedings of COMMA 2014, Atholl Palace Hotel, Scottish Highlands, UK, September 9-12, 2014*, volume 266 of *Frontiers in Artificial Intelligence and Applications*, pages 421–432. IOS Press, 2014.

[60] Wolf Vanpaemel, Gerrit Storms, and Bart Ons. A varying abstraction model for categorization. In Bruno G. Bara, Lawrence Barsalou, and Monica Bucciarelli, editors, *Proceedings of the 27th Annual Conference of the Cognitive Science Society*, pages 2277–2282, Mahwah, NJ, 2005. Lawrence Erlbaum.

[61] Bart Verheij. Dialectical argumentation with argumentation schemes: An approach to legal logic. *Artif. Intell. Law*, 11(2-3):167–195, 2003.

[62] Hajime Yoshino. The systematization of legal meta-inference. In L. Thorne McCarty, editor, *Proceedings of the Fifth International Conference on Artificial Intelligence and Law, ICAIL '95, College Park, Maryland, USA, May 21-24, 1995*, pages 266–275. ACM, 1995.

[63] Jia-Huai You, Xianchang Wang, and Li-Yan Yuan. Nonmonotonic reasoning as prioritized argumentation. *IEEE Trans. Knowl. Data Eng.*, 13(6):968–979, 2001.

[64] Anthony P. Young, Sanjay Modgil, and Odinaldo Rodrigues. Prioritised default logic as rational argumentation. In Catholijn M. Jonker, Stacy Marsella, John Thangarajah, and Karl Tuyls, editors, *Proceedings of the 2016 International Conference on Autonomous Agents & Multiagent Systems, Singapore, May 9-13, 2016*, pages 626–634. ACM, 2016.

[65] Lotfi A. Zadeh. Fuzzy sets. *Information and Control*, 8:338–353, 1965.

[66] Tomasz Zurek and Michal Araszkiewicz. Modeling teleological interpretation. In *International Conference on Artificial Intelligence and Law, ICAIL '13, Rome, Italy, June 10-14, 2013*, pages 160–168, 2013.

LOGICS FOR GAMES, EMOTIONS AND INSTITUTIONS

EMILIANO LORINI
IRIT-CNRS, Toulouse University, France
Emiliano.Lorini@irit.fr

Abstract

We give an informal overview of the way logic and game theory have been used in the past and are currently used to model cognitive agents and multi-agent systems (MAS). In the first part of the paper we consider formal models of mental attitudes and emotions, while in the second part we move from mental attitudes to institutions via collective attitudes.

1 Introduction

Agents in the societies can be either human agents or artificial agents. The focus of this paper is both on: (i) the present society in which human agents interact with the support of ICT through social networks and media, and (ii) the future society with mixed interactions between human agents and artificial systems such as autonomous agents and robots. Indeed, new technologies will come for future society in which such artificial systems will play a major role, so that humans will necessarily interact with them in their daily lives. This includes autonomous cars and other vehicles, robotic assistants for rehabilitation and for the elderly, robotic companions for learning support.

There are two main general observations underlying the present paper. The first is that interaction plays a fundamental role in existing information and communication technologies (ICT) and applications (e.g., Facebook, Ebay, peer-to-peer systems) and will become even more fundamental in future ICT. The second is that the cognitive aspect is crucial for the design of intelligent systems that are expected to interact with human agents (e.g., embodied conversational agents, robotic assistants, etc.). The system must be endowed with a psychologically plausible model of reasoning and cognition in order to be able (i) to understand the human agent's needs and to predict her behaviour, and (ii) to behave in a believable way thereby meeting the human agent's expectations.

Formal methods have been widely used in artificial intelligence (AI) and in the area of multi-agent systems (MAS) for modelling intelligent systems as well as different aspects of social interaction between artificial and/or human agents. The aim of the present paper is to offer a general overview of the way logic and game theory have been and can be used in AI in order to build formal models of socio-cognitive, normative and institutional phenomena.

We take a bottom-up perspective to the analysis of normative and institutional facts that is in line with some classical analysis in organization theory such as the one presented in March & Simon's famous book "Organizations" [102], described as a book in which they:

> "...surveyed the literature on organization theory, starting with those theories that viewed the employee as an instrument and physiological automaton, proceeding through theories that were centrally concerned with the motivational and affective aspects of human behavior, and concluding with theories that placed particular emphasis on cognitive processes" [102, p. 5].

The present paper is organized in two main sections. Section 2 is devoted to cognitive aspects, while Section 3 is devoted to institutional ones. Section 2 starts from the assumption that cognitive agents are, by definition, endowed with a variety of mental attitudes such as beliefs, desires, preferences and intentions that provide input for practical reasoning and decision-making, trigger action execution, and generate emotional responses. We first present a conceptual framework that:

- clarifies the relationship between intention and action and the role of intention in practical reasoning;

- explains how moral attitudes such as standards, ideals and moral values influence decision-making;

- explains how preferences are formed on the basis of desires and moral values;

- clarifies the distinction between the concept of goal and the concept of preference;

- elucidates how mental attitudes including beliefs, desires and intentions trigger emotional responses, and how emotions retroactively influence decision-making and mental attitudes by triggering belief revision, desire change and intention reconsideration.

Then, we explain how game theory and logic have been used in order to develop formal models of such cognitive phenomena. We put special emphasis on a specific branch of game theory, called epistemic game theory, and on a specific family of logics, so-called agent logics. The aim of epistemic game theory is to extend the classical game-theoretic framework with mental notions such as the concepts of belief and knowledge, while agent logics are devoted to explain how different types of mental attitudes (e.g., belief, desires, intentions) are related, how they influence decision and action, and how they trigger emotional responses.

Section 3 builds the connection between mental attitudes and institutions passing by the concept of collective attitude. Collectives attitudes such as joint intention, group belief, group goal, collective acceptance and joint commitment have been widely explored in the area of collective intentionality, the domain of social philosophy that studies how agents function and act at the group level and how institutional facts relate with physical (brute) facts (cf. [100; 140] for a general introduction of the research in this area). Section 3 is devoted to explain (i) how collective attitudes such as collective acceptance or common belief are formed either through aggregation of individual attitudes or through a process of joint perception, (ii) how institutional facts are grounded on collective attitudes and, in particular, how the existence and modification of institutional facts depend on the collective acceptance of these facts by the agent in the society and on the evolution of this collective acceptance. We also discuss existing logics for institutions that formalize the connection between collective attitudes and institutional facts.

In Section 4 we conclude by briefly considering the opposite path leading from norms and institutions to minds. In particular, we explain how institutions and norms, whose existence depends on their acceptance by the agents in the society, retroactively influence the agents' mental attitudes, decisions and actions.

2 Mental attitudes and emotions

In this section, we start with a discussion of two issues related with the representation of mental attitudes and emotions: (i) the cognitive processing leading from goal generation to action (Section 2.1), and (ii) the representation of the cognitive structure of emotions and of their influence on behaviour (Section 2.2). Then, we briefly explain how these cognitive aspects have been incorporated into game theory (Section 2.3). Finally, we consider how mental attitudes and emotion are formalized in logic and the connection between the representation of mental attitudes in logic and the representation of mental attitudes in game theory (Section 2.4).

2.1 A cognitive architecture

The conceptual background underlying our view of mental attitudes is summarized in Figure 1. (Cf. [90] for a logical formalization of some aspects of this view.) The cognitive architecture represents the process leading from generation of desires and moral values and formation of beliefs via sensing to action performance.

The origin of beliefs, desires and moral values An important and general distinction in philosophy of mind is between epistemic attitudes and motivational attitudes. This dis-

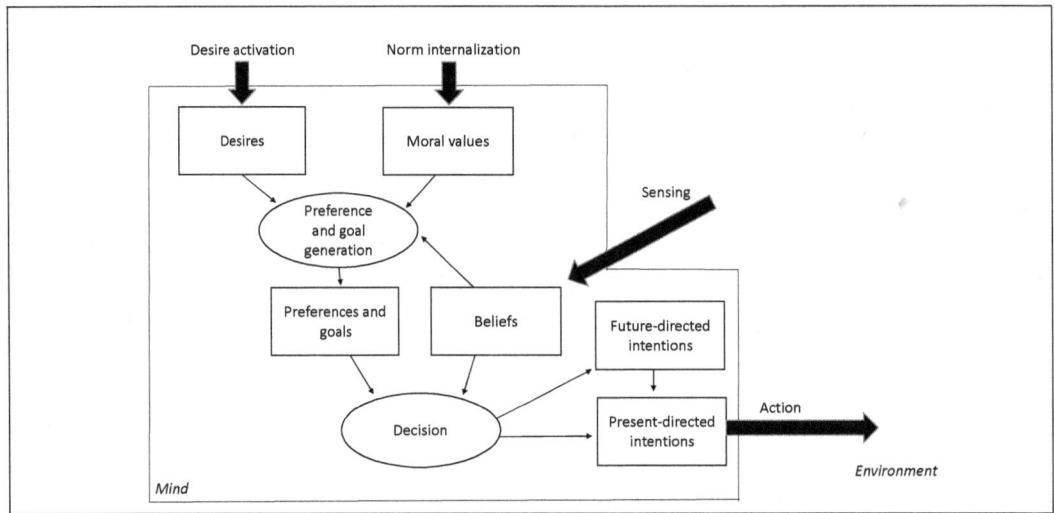

Figure 1: Cognitive architecture

tinction is in terms of the *direction of fit* of mental attitudes to the world. While epistemic attitudes aim at being true and their being true is their fitting the world, motivational attitudes aim at realization and their realization is the world fitting them [114; 7; 67]. Searle [125] calls "mind-to-world" the first kind of *direction of fit* and "world-to-mind" the second one.

There are different kinds of epistemic and motivational attitudes with different functions and properties. Examples of epistemic attitudes are beliefs, knowledge and opinions, while examples of motivational attitudes are desires, preferences, moral values and intentions. However, the most primitive and basic forms of epistemic and motivational attitudes are beliefs, desires and moral values.

Beliefs are mental representations aimed at representing how the physical, mental and social worlds are. Indeed, there are beliefs about natural facts and physical events (e.g., I believe that tomorrow will be a sunny day), introspective beliefs (e.g., I believe that I strongly wish that tomorrow will be a sunny day), and beliefs about mental attitudes of other agents (e.g., I believe that you believe that tomorrow will be a sunny day).

Following the Humean conception, a desire can be viewed as an agent's attitude consisting in an anticipatory mental representation of a pleasant state of affairs (representational dimension of desires) that motivates the agent to achieve it (motivational dimension of desires). The motivational dimension of an agent's desire is realized through its representational dimension, in the sense that, a desire motivates an agent to achieve it *because* the agent's anticipatory representation of the desire's content gives her pleasure so that the agent is "attracted" by it. For example when an agent desires to eat sushi, she is pleased to imagine herself eating sushi. This pleasant representation motivates her to go to the "The Japoyaki" restaurant in order to eat sushi. This view of desires unifies the standard theory of desire (STD) — focused on the motivational dimension — and the hedonic theory of desire (HTD) — focused on the hedonic dimension —. A third theory of desire has been advanced in the philosophical literature (see [124]), the so-called reward theory of desire (RTD). According to RTD what qualifies a mental attitude as a desire is the exercise of a capacity to represent a certain fact as a reward.[1]

Another fundamental aspect of desire is the *longing aspect*. The idea is that for an agent to desire something, the agent should be in a situation in which she does not have what she desires and she yearns for it. In other words, a state of affairs is desired by an agent only if the agent conceives it as *absent*. The following quotation from Locke [86, Book II, Chap. XXI] makes this point clear:

> To return then to the inquiry, what is it that determines the will in regard to our actions? And that...is not, as is generally supposed, the greater good in view: but some (and for the most part the most pressing) uneasiness a man is at present under. This that which successively determines the will, and sets us upon those actions, we perform. This uneasiness we may call, as it is, desire; which is uneasiness of the mind for want of some *absent good*...

This quotation seems in contradiction with what we claimed above, namely, that desire is based on the anticipatory representation of a pleasant state of affairs. However, the stronger the anticipated pleasure associated with a desire, the more painful is its current lack of fulfillment — the term "uneasiness" in the previous quotation —, as in the case of longing for a drink when thirsty, for instance. So the contradiction is only apparent. This aspect of uneasiness described by Locke should not be confused with the concept of aversion which is traditionally opposed to the concept of desire (see [124, Chap. 5]). As emphasized above, if an agent desires a certain fact to be true, then she possesses an anticipatory mental representation of a *pleasant* fact motivating her to make the fact true. On the contrary, if an agent is averse to something, then she possesses an anticipatory mental representation of an *unpleasant* fact motivating her to prevent the fact from being true.

Moral values, and more generally moral attitudes (ideals, standards, etc.), originate from an agent's capability of discerning what from her point of view is (morally) good from what is (morally) bad. If an agent has a certain ideal φ, then she thinks that the realization of the state of affairs φ ought to be promoted because φ is good in itself. Differently from desires, moral values do not necessarily have a hedonic and somatic component: their

---

[1] According to [39], desire is also a necessary condition for reward. In particular, desire determines what counts as a reward for an agent. For example, a person can be rewarded with with water only if she is thirsty and she desires to drink.

fulfillment does not necessarily give pleasure and their transgression does not necessarily give displeasure 'felt' from the body.

There are different ways to explain the origin of beliefs, desires, moral values. Beliefs are formed either via direct sensing from the external environment (e.g., I believe that there is a fire in the house since I can see it), communication (e.g., I believe that there is a fire in the house since you told me this and I trust what you say) and inference (e.g., I believe that there is a fire in the house since I already believe that smoke comes out from the house and if there is smoke coming out from the house then there is fire). One might argue that belief formation via direct sensing is more primitive than belief formation via communication and that the latter can be reduced to the former. Indeed, in the context of communication, the hearer first *perceives* the speaker's utterance, which is nothing but the performance of a physical action (e.g., uttering a certain sound, performing a certain gesture, emitting a certain light signal, etc.) and forms a belief about what the speaker has uttered. Then, she infers the meaning of the speaker's utterance (i.e., what the speaker wants to express by uttering a certain sound, by performing a certain gesture, by emitting a certain light signal, etc.). Although this is true for communication between humans and between artificial systems situated in the physical environment such as robots, it is not necessarily true for communication in an artificial domain in which there is no precise distinction between an utterance and its meaning. In the latter situation, the speaker may transmit to the hearer a message (e.g., a propositional formula) with a precise and non-ambiguous meaning.

The concept of trust plays a fundamental role in belief formation via direct sensing and via communication. Indeed, the hearer will not believe what the speaker says unless she believes that the speaker is a reliable source of information, thereby trusting the speaker's judgment. Similarly, for belief formation via direct sensing, an agent will not believe what she sees unless she believes that her perceptual apparatus works properly, thereby trusting it. The issue whether trust is reducible to other mental attitudes is relevant here. A justifiable approach consists in conceiving *communication-based trust* as a belief about the reliability of a source of information, where "reliable" means that, in the normal conditions, what the source says about a given issue is true.

The explanation about the origin of desires adopted in Figure 1 is that they are activated under certain conditions. For instance, according to Maslow's seminal theory of human motivation, "...everyday conscious desires are to be regarded as symptoms, as surface indicators of more basic needs" [103, p. 392]. Maslow identified a set of basic (most of the time unconscious) needs of human agents including physiological needs,[2] need for safety, need for love and belonging, need for self-esteem and need for self-actualization. For example, a human agent's desire of drinking a glass of water could be activated by her basic

---

[2]Maslow referred to the concept of homeostasis, as the living system's automatic efforts to maintain a constant, normal state of the blood stream, body temperature, and so on.

physiological need for bodily balance including a constant body temperature, constant salt levels in the body, and so on. If certain variables of the agent's body are unbalanced and this unbalance is detected,[3] the agent receives a negative unpleasant signal from her body thereby entering in a state of felt displeasure and uneasiness — in the Lockean sense —. Consequently, she becomes intrinsically motivated to restore bodily balance. The connection between the agent's basic need for bodily balance and the agent's desire of drinking a glass of water may rely on the agent's previous experiences and be the product of operant conditioning (also called instrumental learning). Specifically, the agent may have learnt that, under certain conditions, drinking a glass a water is "a suitable means for" restoring balance of certain variables of the body. Indeed, every time the agent drunk water when she was feeling thirsty, she got a reward by making her basic need for bodily balance satisfied.[4]

In the case of artificial agents, conditions of desire activation should be specified by the system's designer. For example, a robotic assistant who has to take care of an old person could be designed in such a way that, every day at 4 pm, the desire of giving a medicine to the old person is activated in its mind.

As for the origin of moral values, social scientists (e.g., [6]) have defended the idea that there exist innate moral principles in humans such as fairness which are the product of biological evolution. Other moral values, as highlighted in Figure 1, have a cultural and social origin, as they are the product of the internalization of some external norm. A possible explanation is based on the hypothesis that moral judgments are true or false only in relation to and with reference to one or another agreement between people forming a group or a community. More precisely, an agent's moral values are simply norms of the group or community to which the agent belongs that have been internalized by the agent. This is the essence of the philosophical doctrine of moral relativism (see, e.g., [20]). For example, suppose that an agent believes that in a certain group or community there exists a norm (e.g., an obligation) prescribing that a given state of affairs should be true. Moreover, assume that the agent identifies herself as a member of this group or community. In this case, the agent will internalize the norm, that is, the external norm will become a moral value of the agent and will affect the agent's decisions. For example, suppose that a certain person is (and identifies herself as) citizen of a given country. As in every civil country, it is prescribed that citizens should pay taxes. Her sense of national identity will lead the person to adopt the obligation by imposing the imperative to pay taxes to herself. When deciding to pay taxes or not, she will decide to do it, not simply in order to avoid being sanctioned and being exposed to punishment, but also because she is motivated by the moral obligation

---

[3]Converging empirical evidences from neuroscience show that the hypothalamus is responsible for monitoring these bodily conditions.

[4]Following [124], one might argue that most conscious desires (including the desire to eat at a particular time and the desire to drink water) are instrumental, as they are activated *in order to* satisfy more basic needs of the individual.

to paying taxes.

From desires and moral values to preferences According to contemporary theories of human motivation both in philosophy and in economics (e.g., [127; 60]), preferences of a rational agent may originate either (i) from somatically-marked motivations such as desires or physiological needs and drives (e.g., the goal of drinking a glass of water originated from the phisiological drive of thirst), or (ii) from moral considerations and values (e.g., the goal of helping a poor person originated from the moral value of taking care of needy people). More generally, there exists desire-dependent preferences and desire-independent ones originated from moral values. This distinction allows us to identify two different kinds of moral dilemmas. The first kind of moral dilemma is the one which is determined by the logical conflict between two moral values. The paradigmatic example is the situation of a soldier during a war. As a member of the army, the soldier feels obliged to kills his enemies, if this is the only way to defend his country. But, as a catholic, he thinks that human life should be respected. Therefore, he feels morally obliged not to kill other people. The other kind of moral dilemma is the one which is determined by the logical conflict between desires and moral values. The paradigmatic example is that of Adam and Eve in the garden of Eden. They are tempted by the desire to eat the forbidden fruit and, at the same time, they have a moral obligation not to do it.

According to the cognitive architecture represented in Figure 1, desires and moral attitudes of an agent are two different parameters affecting the agent's preferences. This allows us to draw the distinction between *hedonistic* agents and *moral* agents. A purely hedonistic agent is an agent who acts in order to maximize the satisfaction of her own desires, while a purely moral agent is an agent who acts in order to maximize the fulfillment of her own moral values. In other words, if an agent is purely hedonistic, the utility of an action for her coincides with the personal good the agent will obtain by performing this action, where the agent's personal good coincides with the satisfaction of the agent's own desires. If an agent is purely moral, the utility of an action for her coincides with the moral good the agent will promote by performing this action, where the agent's promotion of the moral good coincides with the accomplishment of her own moral values. Utility is just the quantitative counterpart of the concept of preference, that is, the more an agent prefers something, the higher its utility. Of course, purely hedonistic agents and purely moral agents are just extremes cases. An agent is more or less moral depending on whether the utility of a given option for her is more or less affected by her moral values. More precisely, the higher is the influence of the agent's moral values on evaluating the utility of a given decision option, the more moral the agent is. The extent to which an agent's utility is affected by her moral values can be called *degree of moral sensitivity*.[5]

---

[5]This degree can be conceived as a personality trait. In the case of human agents, it is either culturally

Goals The reason why, in Figure 1, preferences and goals are included in the same box is that we conceive goals as intimately related with preferences. In particular, we assume that an agent has φ as a goal (or wants to achieve φ) if and only if: (i) the agent prefers φ to be true to φ to be false, and (ii) the agent considers φ a possible state of affairs (φ is compatible with what the agent believes). The second property is called *realism* of goals by philosophers (cf. [22; 37; 104]). It is based on the idea that an agent cannot reasonably pursue a goal unless she thinks that she can *possibly* achieve it, i.e., there exists at least one possible evolution of the world (a history) that the agent considers possible along which φ is true. Indeed, an agent's goal should not be incompatible with the agent's beliefs. This explains the influence of beliefs on the goal generation process, as depicted in Figure 1.[6] The first property is about the motivational aspect of goals. For φ to be a goal, the agent should not be indifferent between φ and $\neg\varphi$, in the sense that, the agent prefers a situation in which φ is true to a situation in which φ is false, *all other things being equal*. In other words, the utility of a situation increases in the direction by the formula φ *ceteris paribus* ("all else being equal") [154]. This property also defines Von Wright's concept of "preference of φ over $\neg\varphi$" [152].[7] According to this interpretation, a goal is conceived as a *realistic ceteris paribus preference for φ*.

However not all goals have the same status. Certain goals have a motivating force while others do not have it. Indeed, the fact that the agent prefers φ being true to φ being false does not necessarily imply that the agent is motivated to achieve a state in which φ is true and that she decides to perform a certain action *in order to* achieve it. For φ to be a motivating goal, for every possible situation that the agent envisages in which φ is true and for every possible situation that the agent envisages in which φ is false, the agent has to prefer the former to the latter. In other words, there is no way for the agent to be satisfied without achieving φ.[8]

An example better clarifies this point. Suppose Mary wants to buy a reflex camera Nikon and, at the same time, she would like to spend no more than 300 euros. In other words, Mary has two goals in her mind:

- G1: the goal of buying a reflex camera Nikon, and

- G2: the goal of spending no more than 300 euros.

She goes to the shop and it turns out that all reflex cameras Nikon cost more than 300 euros. This implies that Mary believes that she cannot achieve the two goals at the same

---

acquired or genetically determined. In the case of artificial agents, it is configured by the system designer.

[6]The idea that beliefs form an essential ingredient of the goal generation process is also suggested by [26].

[7]Von Wright presents a more general concept of "preference of φ over ψ" which has been recently formalized in a modal logic setting by [146]. See also [119] for an interpretation of this *ceteris paribus* condition based on the concept of logical independence between formulas.

[8]The term 'satisfied' just means that the agent achieves what she prefers.

time, as she envisages four situations in her mind but only three are considered possible by her: the situation in which only the goal G1 is achieved, the situation in which only the goal G2 is achieved and the situation in which no goal is achieved. The situation in which both goals are achieved is considered impossible by Mary. This is not inconsistent with the previous definition of goal since Mary still believes that it is possible to achieve each goal separately from the other. Figure 2 clearly illustrates this: the full rectangle includes all worlds that Mary envisages, so-called *information set*, while the dotted rectangle includes all worlds that Mary considers actually possible, so-called *belief set*.[9] (Cf. [76; 91] for a logical account of the distinction between information set and belief set.)

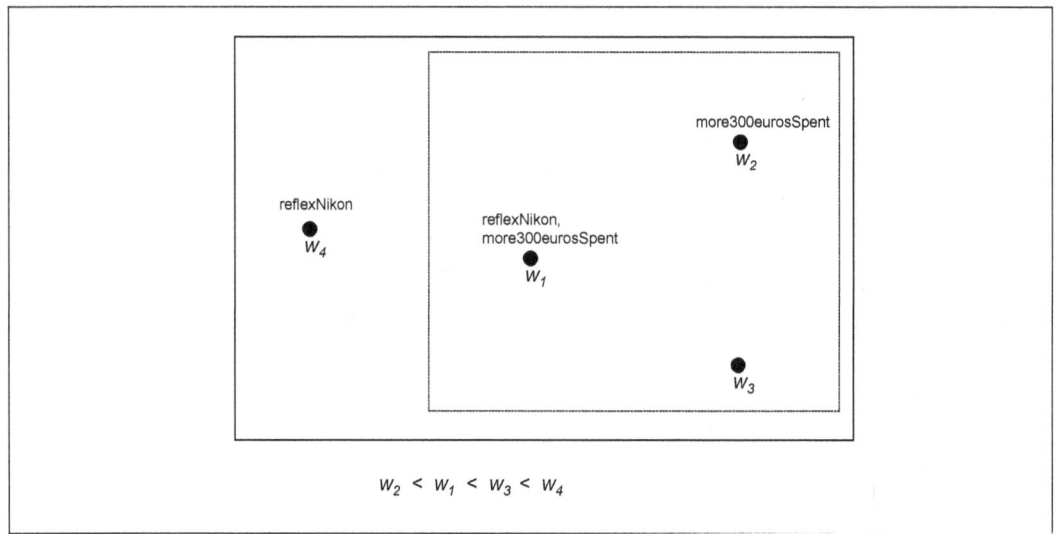

Figure 2: Example for goals

Mary decides to save her money since the goal G2 is a motivating one, while the goal G1 is not. To see that G1 is a goal, it is sufficient to observe that, *all other things being equal*, Mary prefers a situation in which she buys a Nikon to the situation in which she does not buy it. In fact, w_4 is preferred to w_3 and w_1 is preferred to w_2. Moreover, w_4 and w_3 are equal in everything except at w_4 Mary buys a Nikon while at w_3 she does not. Similarly, w_1 and w_2 are equal in everything except at w_1 Mary buys a Nikon while at w_2 she does not. To see that G1 is not motivating, it is sufficient to observe that there exists a situation in which Mary does not buy a Nikon (w_3) that is preferred to a situation in which she does it (w_1). Finally, to see that G2 is a motivating goal, we just need to observe that

---

[9]Mary's information set includes all worlds that, according to Mary, are compatible with the laws of nature. For instance, Mary can perfectly envisage a world in which she is the president of French republic even though she considers this actually impossible.

every situation in which she spends no more than 300 euros (w_3 and w_4) is preferred to every situation in which this is not the case (w_1 and w_2). Thus, on the basis of what she believes, Mary concludes that she can only achieve her goal G2 by saving her money and by buying nothing in the shop.

From preferences and beliefs to actions As the cognitive architecture in Figure 1 highlights, beliefs and preferences are those mental attitudes which determine the agent's choices and are responsible for the formation of new intentions about present actions (present-directed intentions) and future actions (future-directed intentions). As emphasized in the literature in philosophy [22; 105] and AI [23], a future-directed intention is the element of a partial or a complete plan of the agent: an agent may have the intention to perform a sequence of actions later (e.g., the action of going to the train station in two hours followed by the action of taking the train from Paris to Bruxelles at 10 am) in order to achieve a certain goal (e.g., the goal of being in Bruxelles at the European Commission at 2 pm). A present-directed intention is a direct motivation to perform an action now.

In particular, decision is determined by beliefs, preferences and a general rationality criterion stating what an agent should do on the basis of what she believes and what she prefers. Different kinds of rationality criteria have been studied in the areas of decision theory and game theory ranging from expected utility maximization, maxmin and maxmax to satisficing [133]. Once the choice has been made by the agent and the corresponding intention has been formed, the action is performed right afterwards or later. Specifically, an agent forms the intention to perform a certain action at a given point in time and, once the time of the planned action execution is attained, the agent performs the action unless before attaining it, she has reconsidered her prior intention.

2.2 A cognitive view of emotion

In the recent years, emotion has become a central topic in AI. The main motivation of this line of research lies in the possibility of developing computational and formal models of artificial agents who are expected to interact with humans. To ensure the accuracy of a such formal models, it is important to consider how emotions have been defined in the psychological literature. Indeed, in order to build artificial agents with the capability of recognizing the emotions of a human user, of behaving in a believable way, of affecting the user's emotions by the performance of actions directed to her emotions (e.g. actions aimed at reducing the human's stress due to his negative emotions, actions aimed at inducing positive emotions in the human), such agents must be endowed with an adequate model of human emotions.

Appraisal theory The most popular psychological theory of emotion in AI is the so-called appraisal theory (cf. [123] for a broad introduction to the developments in appraisal theory). This theory has emphasized the strong relationship between emotion and cognition, by stating that each emotion can be related to specific patterns of evaluations and interpretations of events, situations or objects (appraisal patterns) based on a number of dimensions or criteria called *appraisal variables* (e.g. goal relevance, desirability, likelihood, causal attribution). Appraisal variables are directly related to the mental attitudes of the individual (e.g. beliefs, predictions, desires, goals, intentions). For instance, when prospecting the possibility of winning a lottery and considering 'I win the lottery' as a desirable event, an agent might feel an intense hope. When prospecting the possibility of catching a disease and considering 'I catch a disease' as an undesirable event, an agent might feel an intense fear.

Most appraisal models of emotions assume that explicit evaluations based on evaluative beliefs (i.e. the belief that a certain event is good or bad, pleasant or unpleasant, dangerous or frustrating) are a necessary constituent of emotional experience. On the other hand, there are some appraisal models mostly promoted by philosophers [126; 50] in which emotions are reduced to specific combinations of beliefs and desires, and in which the link between cognition and emotion is not necessarily mediated by evaluative beliefs. Reisenzein [118] calls *cognitive-evaluative* the former and *cognitive-motivational* the latter kind of models. For example, according to cognitive-motivational models of emotions, a person's happiness about a certain fact φ can be reduced to the person's belief that φ obtains and the person's desire that φ obtains. On the contrary, according to cognitive-evaluative models, a person feels happy about a certain fact φ if she believes that φ obtains and she evaluates φ to be good (desirable) for her. The distinction between cognitive-evaluative models and cognitive-motivational models is reminiscent of the opposition between the Humean view and the anti-Humean view of desire in philosophy of mind. According to the Humean view, belief and desires are distinct mental attitudes that are not reducible one to the other. Moreover, according to this view, there are no necessary connections between beliefs and desires, i.e., beliefs do not necessarily require corresponding desires and, viceversa, desires do not necessarily require corresponding beliefs. On the contrary, the anti-Humean view defends the idea that beliefs and desires are necessarily connected. A specific version of anti-Humeanism is the so-called "Desire-as-Belief Thesis" criticized by the philosopher David Lewis in [80] (see also [81; 57]). In line with cognitive-evaluative models, this thesis states that an agent *desires* something to the extent that she *believes* it to be good.

The popularity of appraisal theory in logic and AI is easily explained by the fact that it perfectly fits with the concepts and level of abstraction of existing logical and computational models of cognitive agents developed in these areas. Especially cognitive-motivational models use folk-psychology concepts such as belief, knowledge, desire and intention that are traditionally used in logic and AI for modelling cognitive agents.

The conceptual background underlying our view of appraisal theory is depicted in Figure 3 which is nothing but the cognitive architecture of Figure 1 extended with an emotion component.

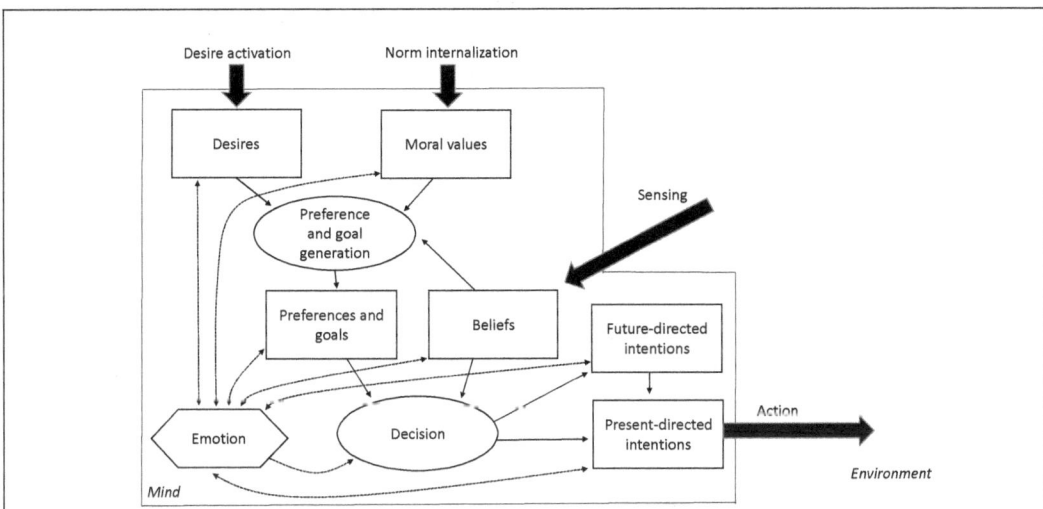

Figure 3: Cognitive architecture extended with emotions

Figure 3 highlights the role of mental attitudes in emotion. In particular, it highlights the fact that mental attitudes of different kinds such as belief, desires, preferences, goals, moral values and (present-directed or future-directed) intentions determine emotional responses. For example, as emphasized above, the emotional response of happiness is triggered by a *goal* and the *certain belief* that the content of one's goal is true. On the contrary, the emotional response of sadness is triggered by a *goal* and the *certain belief* that the content of one's goal is false. The emotional response of hope is triggered by a *goal* and the *uncertain belief* that the content of one's goal is true. On the contrary, the emotional response of fear is triggered by a *goal* and the *uncertain belief* that the content of one's goal is false. This view is consistent with a famous appraisal model, the so-called OCC psychological model of emotions [111], according to which, while joy and distress are triggered by *actual consequences*, hope and fear are triggered by *prospective consequences* (or *prospects*). [52] interpret the term 'prospect' as synonymous of 'uncertain consequence' (in contrast with 'actual consequence' as synonymous of 'certain consequence').

Moral guilt and reproach are examples of emotions that are triggered by moral values [56]. While moral guilt is triggered by the *belief* of being responsible for the violation of a *moral value* or the *belief that one is responsible for having behaved in a morally reprehensible way*, reproach is triggered by the *belief* that someone else is responsible for the violation of a *moral value* or *belief that someone else is responsible for having behaved in a morally reprehensible way.* In other words, guilt is triggered by self-attribution of responsibility for the violation of a moral value, while reproach is triggered by attribution to others of responsibility for the violation of a moral value.

Intentions as well might be responsible for triggering certain kinds of emotional response. For instance, as emphasized by psychological theories of anger (e.g., [77; 111; 121]), a necessary condition for an agent 1 to be angry towards another agent 2 is the agent 1's belief that agent 2 has performed an action that has damaged her, that is, 1 believes that she has been kept from attaining an important goal by an improper action of agent 2. Anger becomes more intense when agent 1 believes that agent 2 has *intentionally* caused the damage. In this sense, an agent 1's belief about another agent 2's intention may have implications on the intensity of agent 1's emotions.

Figure 3 also represents how emotions retroactively influence mental states and decision either (i) through coping or (ii) through anticipation and prospective thinking (i.e., the act of mentally simulating the future) in the decision-making phase.

Coping is the process of dealing with emotion, either externally by forming an intention to act in the world (problem-focused coping) or internally by changing the agent's interpretation of the situation and the mental attitudes that triggered and sustained the emotional response (emotion-focused coping) [77]. For example, when feeling an intense fear due to an unexpected and scaring stimulus, an agent starts to reconsider her beliefs and intentions in order to update her knowledge in the light of the new scaring information and to avoid running into danger (emotion-focused coping). Then, the agent forms an intention to go out of danger (problem-focused coping). Another agent can try to discharge her feeling of guilt for having damaged someone either by forming the intention to repair the damage (problem-focused coping) or by reconsidering the belief about her responsibility for the damage (emotion-focused coping). The coping process as well as its relation with appraisal is illustrated in Figure 4.

Influence of emotion on decision The influence of emotion on decision-making has been widely studied both in psychology and in economics. Rick & Loewenstein [120] distinguish the following three forms of influence:

- **Immediate emotions**: real emotions experienced at the time of decision-making:

  - **Integral influences**: influences from immediate emotions that arise from contemplating the consequences of the decision itself,
  - **Incidental influences**: influences from immediate emotions that arise from factors unrelated to the decision at hand (e.g., the agent's current mood or chronic dispositional affect);

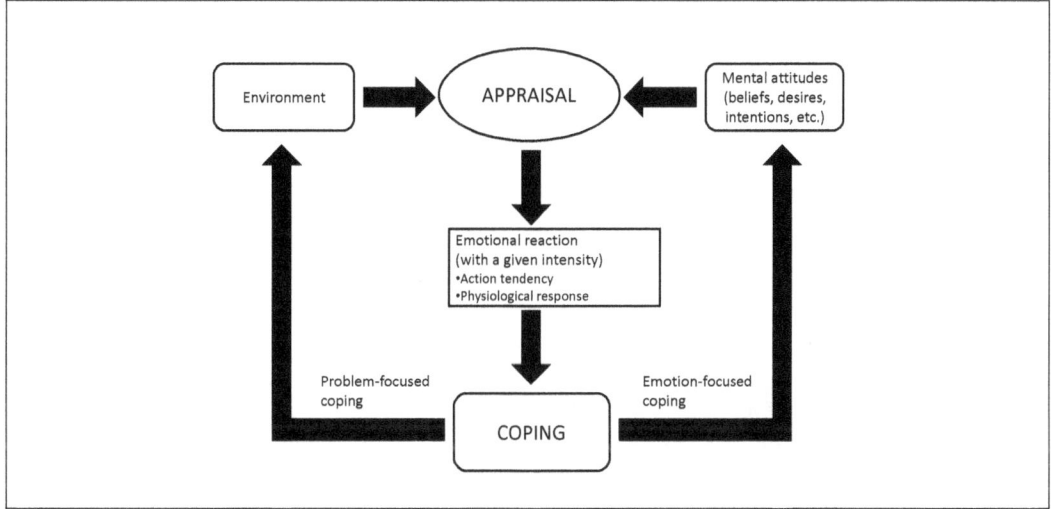

Figure 4: Appraisal and coping cycle

- **Anticipated emotions**: predictions about the emotional consequences of decision outcomes (they are not experienced as emotions per se at the time of decision-making).

An example of integral influence of an immediate emotion is given by the following example.

Example 1. *Paul would like to eat some candies but her mother Mary has forbidden him to eat candies without her permission. Paul's fear of the sanction influences Paul's decision not to eat candies without asking permission.*

The following example illustrates incidental influence of an immediate emotion.

Example 2. *Mary has quarreled with her colleague Paul. At the end of the day she goes back home after work and on the metro a beggar asks her for money. Few hours after the quarrel with Paul, Mary is still in a bad mood and because of her current disposition she refuses the beggar's request.*

The following example illustrates the influence of anticipated emotions on decision.

Example 3. *Peter has to decide whether to leave her job as a researcher at the university of Paris and to accept a job offer as a professor at a university in the U.S. She decides to accept the job offer because she thinks that, if she refuses it, she will likely regret her decision.*

One of the most prominent theory of the integral influence of emotion on decision is Damasio's theory of the somatic marker [35]. According to this theory, decision between different courses of actions leads to potentially advantageous (positive) or harmful (negative) outcomes. These outcomes induce a somatic response used to mark them and to signal their danger or advantage. In particular, a negative somatic marker 'signals' to the agent the fact that a certain course of action should be avoided, while a positive somatic marker provides an incentive to choose a specific course of action. According to Damasio's theory, somatic markers depend on past experiences. Specifically, pain or pleasure experienced as a consequence of an outcome are stored in memory and are felt again when the outcome is envisaged in the decision-making process. The following example clearly illustrates this.[10]

Example 4. *Mary lives in Toulouse and has to decide whether to go to Paris by plane or by train. Last time she traveled by plane she had a painful experience because of turbulence. Mary envisages the possibility of incurring again in a turbulence and gets frightened, thereby deciding to travel by train.*

Several works aimed at extending the classical expected utility model to incorporate anticipated emotions that are related to our uncertainty about the future, such as hopefulness, anxiety, and suspense [27]. Some economic models of decision-making consider how the anticipation of a future regret might affect a person's current decision [87]. In particular, according to these models, if a person believes that after choosing a certain action she will likely regret for having made this choice, she will be less willing to choose the action (than in the case in which she does not believe this). These models agree in defining regret as the emotion that stems from the comparison between the actual outcome deriving from a given choice and a counterfactual better outcome that might have been had one chosen a different action [45; 70; 157]. More recently, some economists have studied the influence of strategic emotions such as interpersonal guilt and anger on decision [10; 29; 65]. Following psychological theories of interpersonal guilt [12; 139], models developed in this area assume that the prototypical cause of guilt is the infliction of harm, loss, or distress on a relationship partner. Moreover, they assume that if people feel guilty for hurting their partners and for failing to live up to their expectations, they will alter their behavior (to avoid guilt) in ways that seem likely to maintain and strengthen the relationship. This is different from the concept of moral guilt formalized by [97] according to which a person feels (morally) guilty if she believes that she is responsible for having behaved in a morally reprehensible way (see Section 2.4 for more details).

---

[10]Positive and negative somatic markers can operate either at a conscious level or at a unconscious/automatic level. This corresponds to Ledoux's distinction between explicit memory and implicit memory and between two possible elaborations of a stimulus inducing an emotional response [78]: conscious elaboration vs. automatic elaboration.

2.3 Interacting minds: from game theory to epistemic game theory

The idea highlighted in Section 2.1 of describing rational agents in terms of their epistemic and motivational attitudes, is also adopted by classical decision theory and game theory. In particular, classical decision theory accounts for the criteria and principles (e.g., expected utility maximization) that a rational agent should apply in order to decide what to do on the basis of her beliefs and preferences. Game theory generalizes decision theory to the multi-agent case in which agents' decisions are interdependent and agents' actions might interfere between them so that: (i) the possibility for an agent to achieve her goals may depend on what the other agents decide to do, and (ii) agents form beliefs about the future choices of the other players and, consequently, their current decisions are influenced by what they believe the others will do. More generally, game theory involves a strategic component that is not considered by classical decision theory whose object of analysis is a single agent who makes decisions and acts in an environment she does not share with other agents.

Classical decision theory and game theory provide a quantitative account of individual and strategic decision-making by assuming that agents' beliefs and preferences can be respectively modeled by subjective probabilities and utilities. In particular, while subjective probability captures the extent to which a fact is *believed* by a certain agent, utility captures how much a certain state of affairs is *preferred* by the agent. In other words, subjective probability is the quantitative counterpart of the concept of belief, while utility is the quantitative counterpart of the concept of preference.[11]

One of the fundamental concepts of game theory is the concept of solution which is, at the same time, a prescriptive notion, in the sense that it prescribes how rational agents in a given interaction *should* play, and a predictive one, in the sense that it allows us to predict how the agents *will* play. There exist many different solution concepts both for games in normal form and for games in extensive form (e.g., Nash Equilibrium, iterated deletion of strongly dominated strategies, iterated deletion of weakly dominated strategies, correlated equilibrium, backward induction, forward induction, etc.) and new ones have been proposed in the recent years (see, e.g., [58]). A major issue we face when we want to use a solution concept in order either to predict human behavior or to build some practical applications (e.g., for computer security or for multi-agent systems) is to evaluate its significance. Some of the questions that arise in these situations are, for instance: given certain assumptions about the agents such as the assumption that they are rational (e.g., utility

---

[11]Qualitative approaches to individual and strategic decision-making have been proposed in AI [18; 68] to characterize criteria that a rational agent should adopt for making decisions when she cannot build a probability distribution over the set of possible events and her preference over the set of possible outcomes cannot be expressed by a utility function but only by a qualitative ordering over the outcomes. For example, going beyond expected utility maximization, qualitative criteria such as the maxmin principle (choose the action that will minimize potential loss) and the maxmax principle (choose the action that will maximize potential gain) have been studied and axiomatically characterized [19; 20].

maximizers), under which conditions will the agents converge to equilibrium? Are these conditions realistic? Are they too strong for the domain of application under consideration? There is a branch of game theory, called epistemic game theory, which can help to answer these questions (cf. [113] for a general introduction to the research in this area). Indeed, the aim of epistemic game theory is to provide an analysis of the necessary and/or sufficient epistemic conditions of the different solution concepts, that is, the assumptions about the epistemic states of the players that are necessary and/or sufficient to ensure that they will play according to the prescription of the solution concept. Typical epistemic conditions which have been considered are, for example, the assumption that players have common belief (or common knowledge) about the rationality of every player,[12] the assumption that every player knows the choices of the others,[13] or the assumption that players are logically omniscient.[14]

Epistemic game theory shares concepts and methods with what Aumann calls interactive epistemology [8]. The latter is the research area in logic and philosophy which deals with formal models of knowledge and belief when there is more than one rational agent or "player" in the context of interaction having not only knowledge and beliefs about substantive matters, but also knowledge and beliefs about the others' knowledge and beliefs. The concept of rationality corresponds either to the optimality criterion according to which an agent should choose an action which guarantees the highest utility, given what she believes the other agents will do, or the prudential criterion according to which an agent should not choose an action which ensures the lowest utility, given what she believes the other agents will do. An example of the former is expected utility maximization, while an example of the latter is weak rationality in the sense of [145] (cf. also [109; 15]), according to which an agent should not choose an action which is strongly dominated by another action, given what the agent believes the other agents will do.

Epistemic game theory provides a useful framework for clarifying how agents' mental attitudes influence behaviours of agents in a social setting. In particular, it allows us to understand the subtle connection between beliefs, preferences and decision, as represented in Figure 1 given in Section 2.1, under the assumption that the agents' decisions are interdependent, in the sense that they are affected by what the agents believe the others will choose.[15]

---

[12]This is the typical condition of iterated deletion of strongly dominated strategies (also called iterated strong dominance).

[13]This condition is required in order to ensure that the agents will converge to a Nash equilibrium.

[14]See [158] for an analysis of iterated strong dominance after relaxing the assumption of logical omniscience.

[15]Although epistemic game theory and, more generally, game theory share with Figure 1 the concepts of belief and preference, they do not provide an account of the origin of beliefs, desires and moral values and of the connection between desires, moral values and preferences. Moreover, the concept of future-directed intention is not included in the conceptual apparatus of game theory and epistemic game theory. The same can be said

2.4 Logics for mental attitudes, emotion and games

This section is devoted to discuss existing logics for mental attitudes and emotion proposed in AI as well as the connection between the representation of mental attitudes and emotion in logic and the representation of mental attitudes and emotion in game theory.

Logics for mental attitudes Since the seminal work of [31] aimed at implementing Bratman's philosophical theory of intention [22], many formal logics for reasoning about mental attitudes of agents such as beliefs, desires and intentions have been developed. Among them we should mention the logics developed by [93; 90; 63; 75; 107; 108; 117; 130; 134; 148; 155].

The general term used to refer to this family of logics is *agent logics*. A subfamily is the family of BDI logics whose most representative example is the modal logic by [117] whose primitive constituents are the the concepts of belief (B), desire (D) and intention (I) which are expressed by corresponding modal operators. Another well-known agent logic is the so-called KARO framework developed by [107]. KARO is a multi-modal logic framework based on a blend of dynamic logic with epistemic logic, enriched with modal operators for modeling mental attitudes such as beliefs, desires, wishes, goals and intentions.

Generally speaking, agent logics are nothing but formal models of rational agency whose aim is to explain how an agent endowed with mental attitudes makes decisions on the basis of what she believes and of what she wants or prefers. In this sense, the decisions of the agent are determined by both the agent's beliefs (the agent's epistemic states) and the agent's preferences (the agent's motivational states). As discussed in Section 2.1, the output of the agent's decision-making process is either a choice about what to do in the present, also called present-directed intention, or a choice about what to do in the future, also called future-directed intention. The idea that the behavior of an agent can be explained by attributing mental states to the agent and by having a sophisticated account of the relationship between her epistemic states and her motivational states and of the influence of these on the agent's decision-making process is something agent logics share with other disciplines including philosophy of mind [38], cognitive sciences [116], psychology [118] and artificial intelligence [28].

Logics for emotion More recently, agent logics have been used to formalize the cognitive structure and the coping strategies of different types of emotion. For instance, a logical formalization of emotion in the context of the KARO framework has been proposed. In particular, in the KARO framework each emotion type is represented with a special predicate, or fluent, in the jargon of reasoning about action and change, to indicate that these

---

for goals: the concept of goal is somehow implicit in the utility function but is not explicitly modeled.

predicates change over time. For every fluent a set of effects of the corresponding emotions on the agent's planning strategies are specified, as well as the preconditions for triggering the emotion in terms of mental attitudes of agents. The latter correspond to generation rules for emotions. For instance, in [106] generation rules for four basic emotions are given: joy, sadness, anger and fear, depending on the agent's plans. In [144] generation rules for guilt and shame have been proposed.

A logical formalization of the OCC psychological model of emotions [111] has been proposed in [1].

Surprise is the simplest emotion that is triggered by the mismatch between an expectation that an event will possibly occur and an incoming input (i.e., what an agent perceives). In [92] a logical theory of surprise is proposed. The theory clarifies two important aspects of this cognitive phenomenon. First, it addresses the distinction between surprise and astonishment, the latter being the emotion triggered by something an agent could not reasonably expect. The crucial difference between surprise and astonishment is that the former necessarily requires an explicit expectation in the agent's mind, while the latter does not. One can be astonished by something since, at the moment she perceives it, she realizes that it was totally unpredictable, without having formulated an expectation in advance. For example, suppose Mary is working in her office. Suddenly, someone knocks the door and enters into Mary's office. Mary sees that the person is a policeman. She is astonished by this fact even though, before perceiving it, she did not have explicit in her mind the expectation that "a policeman will not enter into the office". Secondly, the theory clarifies the role of surprise in belief change by conceiving it as a basic mechanism which is responsible for triggering belief reconsideration.

In a more recent paper [99], a logical formalization of counterfactual emotions has been provided. Counterfactual emotions, whose prototypical example is regret, are those emotions that are based on counterfactual reasoning about agents' choices. Other examples are rejoicing, disappointment, and elation. The formalization is based on an epistemic extension of STIT logic (the logic of "seeing to it that") by Belnap et al. [13; 66; 25; 89] and allows to capture the cognitive structure of regret and, in particular, the counterfactual belief which is responsible for triggering this emotion, namely the *belief that a counterfactual better outcome might have been, had the agent chosen a different action*. In [96], the STIT logical analysis of counterfactual emotions is extended to moral emotions. The latter involve counterfactual reasoning about responsibility for the transgression of moral values. In particular, the proposed formalization accounts for the attribution of responsibility for the violation of a moral value either to the self or to the other. This is a fundamental constituent of moral emotions such as guilt, reproach, moral pride and moral approval. For example, according to the proposed analysis, guilt is triggered by the *belief that one is responsible for having behaved in a morally reprehensible way*. A game-theoretic account of moral guilt, which parallels the STIT logical analysis, has been given in [97].

The problem of emotion intensity has also been adressed by logicians. Following existing psychological models of emotion based on appraisal theory, intensity of these emotions is defined as a function of two cognitive parameters, the strength of the expectation and the strength of the desire which are responsible for triggering the emotional response. For instance, the intensity of hope that a certain event will occur is a monotonically increasing function of both the strength of the expectation and the strength of the desire that the event will occur. The logical theory of appraisal and coping presented in [36] also considers the behavioral aspects of such emotions: how the execution of a certain coping strategy depends on the intensity of the emotion generating it. Specifically, it is assumed that: (i) an agent is identified with a numerical value which defines her tolerance to the negative emotion, and (ii) if the intensity of the negative emotion (e.g., fear) exceeds this value then the agent will execute a coping strategy aimed at discharging the negative emotion.

Logics for games The relationship between logic and game theory has been explored in both directions: *games for logic* and *logic for games*. On the one hand, methods and techniques from game theory have been applied to formal semantics, proof theory and model checking for different kinds of logic [64; 51; 72]. On the other hand, logical representation languages have been proposed in computer science and AI to represent game-theoretic concepts such as the concepts of strategy, capability, winning strategy as well as solution concepts such as Nash equilibrium and backward induction. This includes logics such as Coalition Logic [112], Alternating-time Temporal Logic (ATL) [5] and STIT (the logic of "seeing to it that") [13; 66].

More recently, logics for epistemic game theory have been proposed by incorporating epistemic components in existing logics for games and developing new logical formalisms that can represent, at the same time, the structure of the game and the mental attitudes and rationality of the players involved in the game.

Much of the work in the field of epistemic game theory is based on a *quantitative* representation of uncertainty and epistemic attitudes. Notable examples are the analysis of the epistemic foundations for forward induction and for iterated admissibility based on Bayesian probabilities [135; 59], conditional probabilities [11] or lexicographic probabilities [21]. The distinction between quantitative and qualitative approaches to uncertainty has been widely discussed in the AI literature (cf. [49]). While in quantitative approaches belief states are characterized by classical probabilistic measures or by alternative numerical accounts, such as lexicographic probabilities or conditional probabilities [11], qualitative approaches do not use any numerical representation of uncertainty but simply a plausibility ordering on possible worlds structures inducing an epistemic-entrenchment-like ordering on propositions.

Both logics for epistemic game theory based on a qualitative representation of epistemic attitudes [9; 91; 98] and logics for epistemic game theory based on probability theory [59; 17] have been proposed in the recent years. The main motivation for the latter is to exploit logical methods in order to provide sound and complete axiomatics for important concepts studied in epistemic game theory such as rationality and common knowledge of rationality. The main motivation for the former is to show that interesting results about the epistemic foundation for solution concepts in game theory can be proved in a qualitative setting, without necessarily exploiting the complex machinery of probability theory.

The connection between logical models of epistemic states based on Kripke semantics and formal models of epistemic states based on the concept of type space has also been explored [46; 73]. While the former have been mainly proposed by logicians in AI [44] and philosophy [137], the latter have been proposed by game theorists in economics [61]. The main motivation for this research lies in the possibility of building a bridge between two research communities that study the same concepts and phenomena from different perspectives.

3 From mental attitudes to institutions via collective attitudes

In this section we gradually move from minds to institutions. The connection between the former and the latter is built via the concept of collective attitude. Specifically, we discuss a particular view of institutions: the idea that institutional facts are grounded on the agents' collective attitudes that, in turn, originate from the agents' mental attitudes.

Section 3.1 starts with a discussion about the different functions and origins of collective attitudes, while Section 3.2 clarifies the connection between collective attitudes and institutions. Finally, Section 3.3 explains how this connection has been formalized in logic.

3.1 Collective attitudes

Collectives such as groups, teams, coorporations, organizations, etc. do not have minds. However, we frequently ascribe intentional attitudes to them in the same way as we ascribe intentional attitudes to individuals. For example, we may speak of what our family prefers, of what the goal of a coorporation or organization is, of what the scientific community think about a certain issue, and so on.

Aggregate vs. common attitudes An important distinction in the theory of collective attitudes is between aggregate attitudes and common attitudes. As emphasized by [84] "...an aggregate attitude (of a collective) is an aggregate or summary of the attitudes of the individual members of the collective, produced by some aggregation rule or statistical criterion...". A typical example of aggregate attitude produced by a statistical criterion is shared belief, namely the fact that all agents (or most of the agents) in a set of agents believe that a certain proposition p is true. An example of aggregate attitude produced by an aggregation rule is the collective acceptance of a jury about a given proposition p obtained by majority voting: the jury believes that the proposition p is true if and only if the majority of the members of the jury has expressed the individual opinion that p is true. Aggregate attitudes produced by aggregation rules are the objects of analysis of judgement aggregation, an important research area in social sciences and AI (see [54; 83] for an introduction to judgement aggregation). Differently from common attitudes, aggregate attitudes do not require a level of common awareness by the members of the group. That is, a group can hold an aggregate attitude even though the members of the group do not necessarily believe so. For example, the fact that two agents share the belief that p is true does not necessarily imply that they individually believe that they share this belief. As emphasized by [84] "...a common attitude (of a collective) is an attitude held by all individual members of the collective, where their holding it is a matter of common awareness", where the term "common awareness" refers to the fact that every member of the group believes that the group has the common attitude, that every member of the group believes that every member of the group believes that the group has the common attitude, and so on. A typical example of common attitude is common belief: every agent in the group believes that p is true, every agent in the group believes that every agent in the group believes that p is true, and so on ad infinitum.

Functions of collective attitudes Collective attitudes play a crucial role in the society as: (i) they provide the basis of our common understanding through communication, (ii) they ensure coordination between agents, (iii) they are fundamental constituents of collaborative activities between agents acting as members of the same team.

In linguistic, the concept of common ground in a conversation is typically conceived as the common knowledge (or common belief) that the speaker and the hearer have about the rules of the language they use and about the meaning of the expressions uttered by the speaker [136]. Indeed, language use in conversation is a form of social activity that requires a certain level of coordination between what the speaker means and what the addressee understands the speaker to mean. Any utterance of the speaker is in principle ambiguous because the speaker could use it to express a variety of possible meanings. Common ground — as a mass of information and facts mutually believed by the speaker and the addressee — ensures coordination by disambiguating the meaning of the speaker's utterance. For example, suppose two different operas, "Don Giovanni" by Mozart and "Il Barbiere di Siviglia" by Rossini, are performed in the same evening at two different theaters. Mike goes to see Don Giovanni and the next morning sees Mary and asks "Did you enjoy the opera yesterday night?", identifying the referent of the word "opera" as Don Giovanni. In order to ensure that Mary will take "opera" as referring to Don Giovanni, it has to be the case that the night before Mary too went to see Don Giovanni, that Mary believes that Mike too went to see Don Giovanni, that Mary believes that Mike believes that Mary too went to see Don Giovanni, and so on.

Moreover, since the seminal work by David Lewis [82], the concept of common belief has been show to play a central role in the formation and emergence of social conventions.

Finally, collective attitudes such as common goal and joint intention are traditionally used in in the philosophical area and in AI to account for the concept of collaborative activity [24; 55; 40; 41]. Notable examples of collaborative activity are the activities of painting a house together, dancing together a tango, or moving a heavy object together. Two or more agents acting together in a collaborative way need to have a common goal and need to form a shared plan aimed at achieving the common goal. In order to make collaboration effective, each agent has to commit to her part in the shared plan and form the corresponding intention to perform her part of the plan. Moreover, she has to monitor the behaviors of the others and, eventually, to reconsider her plan and adapt her behavior to the new circumstances.

The origin of collective attitudes Where do collective attitudes come from? How are they formed? There is no single answer to these questions, as collective attitudes can originate in many different ways.

As explained above, aggregate attitudes are the product of aggregation procedures like majority voting or unanimity (cf. [85]). The agents in a certain group decide to use a certain aggregation rule. Then, every agent expresses her opinion about a certain issue p and the aggregation rule is used to determine what the group believes or what the group accepts. Examples of collective attitudes originating from the aggregation of individual attitudes are group belief and collective acceptance.

Collective attitudes, such as shared belief and common belief, can also be formed through communication or joint perception. A source of information announces to all agents in a group that a certain proposition p is true. Under the assumption that every agent perceives what the information source says and that every agent in the group trusts the information source's jugement about p, the agents will share the belief that p is true as a result of the announcement. Creation of common belief through communication requires satisfaction of certain conditions that are implicit in the concept of public announcement, as defined in the context of public announcement logic (PAL) [115], the simplest logic in the family of dynamic epistemic logics (DEL) [147]. Specifically, to ensure that an announcement will determine a common belief that the announced fact is true, every agent in the group has to perceive what the information source says, every agent in the group has to perceive that every agent in the group perceives what the information source says, and so on. The latter is called *co-presence* condition in the linguistic literature [30].

The concept of co-presence becomes particularly relevant in the perspective of designing artificial systems situated in a physical environment that need to acquire common belief of certain facts in order to achieve coordination and to make collaboration effective. For example, imagine two robots moving in the physical environment. A source of information signals to them that there is a danger. It does this by emitting a red light. The robots will be able to form different levels of mutual belief about this fact depending on: (i) their spatial positions and the orientation of their sensors with respect to the source of information, and (ii) the perception of the other robots' spatial positions and of the orientations of the other robots' sensors with respect to the source of information. The concept of co-presence applies not only to agents interacting in a physical environment but also to agents interacting in a virtual environment (e.g., virtual characters of a videogame).

A side note: collective acceptance vs. common belief A property that clearly distinguishes collective acceptance from common belief is that common belief implies shared belief, while collective acceptance does not: when there is a common belief in a group of agents C that a certain proposition p is true then each agent in C individually believes that p is true, while it might be the case that there is a collective acceptance in C that p is true, and at the same time one or several agents in C do not individually believe that p is true. For example, the members of a Parliament might collectively accept (*qua* members of the Parliament) that launching a military action against another country is legitimate because by majority voting the Parliament decided so, even though some of them — who voted against the military intervention — individually believe the contrary. This difference is due to the fact that collective acceptance is a kind of aggregate attitude which can be formed through aggregation procedures others than unanimity.

Another important difference between collective acceptance and common belief is the irreducibility of collective acceptance to the individual level. In particular, it has been emphasized that, while common belief is strongly linked to individual beliefs and can be reduced to them, collective attitudes such as collective acceptance cannot be reduced to a composition of individual attitudes. This aspect is particularly emphasized by Gilbert [47] who follows Durkheim's non-reductionist view of collective attitudes [42]. According to Gilbert, any proper group attitude cannot be defined only as a label on a particular configuration of individual attitudes, as common belief is. In [48; 143] it is suggested that a collective acceptance of a set of agents C is based on the fact that the agents in C identify themselves as members of a certain group, institution, team, organization, etc. and recognize each other as members of the same group, institution, team, organization, etc. Common belief and common knowledge, as traditionally defined in epistemic logic [44], do not entail this aspect of mutual recognition and identification with respect to the same group, institution, team, organization, etc.

3.2 Grounding institutions and norms on collective attitudes

In the previous section we have explained how collective attitudes are generated from mental attitudes through aggregation procedures, communication or joint perception.

The next step in our analysis is to explain how institutions and norms are grounded on collective attitudes of different types including collective acceptance and common belief. The term "grounded" means that the existence and the evolution of institutions and norms depend on the existence and the evolution of the collective attitudes of the agents who are members of the institution and who are subject to the norm.

We focus here on two forms of grounding that have been considered in the literature: the grounding of institutions on collective acceptance and the grounding of conventions on common belief.

Collective acceptance and institutions The problem of understanding what institutions are and how they function has been addressed both in social sciences, in philosophy and in legal theory. Computer scientists working in the area of multi-agent systems have been interested in devising artificial institutions, modeling their dynamics and the different kinds of rules and norms of an institution that agents have to deal with. Following [110, p. 3], artificial institutions can be conceived as "the rules of the game in a society or the humanly devised constraints that structure agents' interaction". In some models of artificial institutions norms are conceived as means to achieve coordination among agents and agents are supposed to comply with them and to obey the authorities of the system [43]. More sophisticated models of institutions leave to the agents' autonomy the decision whether to comply or not with the specified rules and norms of the institution [2; 88]. However, all previous models abstract away from the legislative source of the norms of an institution, and from how institutions are created, maintained and changed by their members.

What these models of artificial institutions neglect is the fundamental relationship between institutions and the collective attitudes of their members and, in particular, the fact that the existence and the dynamics of an institution (norms, rules, institutional facts, etc.) are determined by the collective attitudes of the agents which identify themselves as members of the institution. This aspect is emphasized in the following quote from [101, p. 77]:

> "only because institutions are anchored in peoples minds do they ever become behaviorally relevant. The *elucidation of the internal aspect is the crucial step* in adequately explaining the emergence, evolution, and effects of institutions." [Emphasis added].

Prominent philosophical theories of institutional reality conceives collective acceptance as the collective attitude on which institutions are grounded [128; 142]. The relationship between acceptance and institutions has also been emphasized in the philosophical doctrine of Legal Positivism [62]. According to Hart, the foundations of an institution consist of adherence to, or acceptance of, an ultimate rule of recognition by which the validity of any rule of the institution may be evaluated. [16]

Common belief and conventions Convention is a concept that has been widely studied in economics [138], philosophy [16; 141] and computer science [153; 151; 131; 129], given the fundamental role it plays in the regulation of both human and artificial societies.

Eating manners, the kind of clothes we wear in office, and the side of the road on which we drive are mundane examples of convention. Roughly, a social convention is a customary, arbitrary and self-enforcing rule of behavior that is generally followed and expected to be followed in a group or in a society at large [82]. When a social convention is established, everybody behaves in an agreed-upon way even if they did not in fact explicitly agree to behave in this way. A social convention can thus be seen as a kind of tacit agreement that has evolved out of a history of previous interactions [138; 141].

Since the seminal contribution by David Lewis [82], the modern approach to conventions is rooted both in epistemic logic and in evolutionary game theory. The *epistemic approach* to the study of conventions has focused on the characterization of the kind of mutual beliefs and expectations that are required for a group to adopt a certain convention [34; 132; 150] and on the distinction between the epistemic conditions of conventions in contrast with the epistemic conditions of social norms [14]. The epistemic approach clearly highlights the fact that conventions are grounded on collective attitudes. Indeed, according to the well-known definition of convention by David Lewis [82, pp. 76], a given regularity of behavior R is a convention for a population of agents P at a recurrent situation S, only if the agents in the population P *mutually expect* everyone in P to conform to the regularity R in the situation S (and commonly believe so). In other words, for a convention to exist, the agents in the population have to form a mutual expectation about each other's behavior (and a common belief about this). Consider the example of driving on the left-hand side in the UK. This is a convention as every person in the UK expects other people in the UK to drive on the left-hand side of the road. Moreover, every person in the UK expects other people to drive on the left-hand side of the road *because and as long as* she expects other people to expect everyone to drive on the left-hand side of the road.

The *evolutionary approach* to the study of conventions has focused on the conditions under which a certain convention can emerge on a given population of agents depending on the agents' learning capabilities. Notable examples of this approach are the models by Kandori et al. [71] and Young [156] which make predictions about the conditions under which

---

[16] In Hart's theory, the rule of recognition is the rule that specifies the ultimate criteria of validity in a legal system.

agents converge to equilibrium in a certain coordination game by learning the others' play and adjusting their strategies over time. For instance, Kandori et al.'s model investigates the dynamic process that leads the agents to converge to the risk dominant equilibrium in a repeated 2×2 coordination game.

It is worth noting that the epistemic approach and the evolutionary approach to the study of conventions have not yet been reconciled. Indeed, none of the existing evolutionary models of conventions deals with the epistemic aspect of conventions, as they do not assume agents to be cognitive and only consider a simplified notion of convention as a mere regularity of behavior.

3.3 Logics for institutions

In [95] a modal logic of collective acceptance is proposed, in accordance with the philosophical theories of this notion discussed in Section 3.2. In the logic, collective acceptance is conceived as the collective attitude that some agents have *qua* members of the same institution. In particular, a collective acceptance held by a set of agents C *qua* members of a certain institution x is the kind of acceptance the agents in C are committed to when they are "functioning together as members of the institution x", that is, when the agents in C identify and recognize each other as members of the institution x. For example, in the context of the institution Greenpeace agents (collectively) accept that their mission is to protect the Earth *qua* members of Greenpeace. The state of acceptance *qua* members of Greenpeace is the kind of acceptance these agents are committed to when they are functioning together as members of Greenpeace, that is, when they identify and recognize each other as members of Greenpeace. The logic accounts for different kinds of aggregation procedures that the members of an institution may adopt in order to build a collective acceptance of a given fact. This includes unanimity, majority and a criterion based on leadership according to which what the members of an institution collectively accept coincides with the acceptance of the legislator of the institution. Moreover, the logic clearly distinguishes collective acceptance from common belief, by emphasizing the fact that, while common belief is reducible to individual beliefs, collective acceptance cannot be reduced to individual attitudes of the members of an institution. The fact that collective acceptance is not reducible to individual attitudes is reflected in the formal semantics of the logic. While in epistemic logic common belief is commonly represented by means of the transitive closure of the union of the accessibility relations for the individual beliefs, the accessibility relation for collective acceptance is not definable in terms of the accessibility relations for individual beliefs or individual acceptances. Moreover, collective acceptance entails the notion of "group identification" that is not reducible to the individual level.

Following the idea of some prominent philosophical theories of institutions [128; 142] according to which institutional reality only exists in relation with the collective acceptance of institutional facts by the members of the institution, a systematic analysis of institutional concepts in the context of this logic is given. This includes the concepts of weak permission, strong permission, obligation and constitutive rule.

The relationship between the logic of collective of acceptance and existing logics of institutions has also been investigated. This includes the comparison between the logic of collective acceptance and the logic of institutional facts proposed by [69] and refined more recently by [53]. According to [69; 53], the primary aspect of institutional facts is their being true in the context of an institution x.

In [95], the bridge between collective acceptance and informal institutions is built by assuming that:

> a certain fact φ is true in the context of an informal institution x if only if the members of the informal institution x collectively accept that φ is true (in the context of x).

Differently from formal or legal institutions, informal institutions have no official of the law who is in charge of promulgating new norms and who is the guarantor of their validity. An example of informal institution is a language whose rule specifying the relationship between a certain utterance and its meaning is shared by a group of people: in the context of this group, the utterance has a certain meaning since the language speakers collectively accept this.

In [94], the analysis is extended to formal and legal institutions in which legislators and officials of the law exist who are in charge of either creating new norms or suppressing existing ones out of collective deliberation and who are guarantors of the norms' validity. Specifically, it is assumed that:

> a certain fact φ is true in the context of a formal institution x if only if the legislators of the institution x collectively accept that φ is true (in the context of x).

For example, according to the French law, the legal drinking age is 18 since this fact is accepted by the French legal authority. As emphasized in Section 3.2, this is close to Hart's idea that a legal norm exists because it adheres to the standards of validity specified by the ultimate rule of recognition that has to be accepted by the legal authority. For example, the Italian legal authority accepts that a norm is valid as far as it has been promulgated by the Italian parliament and published in the "Gazzetta Ufficiale della Repubblica Italiana" (Official Gazette of the Italian Republic).

4 Conclusion: closing the circle

In the previous sections we have explained: (i) the role of mental attitudes in decision-making and in action performance as well as the relationship between mental attitudes and emotion (Section 2), (ii) how collective attitudes are generated from mental attitudes as well as the relationship between institutions and norms, on the one hand, and collective attitudes, on the other hand (Section 3). More generally, we have moved from the mental level to the collective level and, then, from the collective level to the institutional-normative level. It is now time to close the circle by going back to mind.

The relevant question here is the following: how do institutions and norms, that are grounded on agents' collective attitudes retroactively influence decision-making and action?

First of all, for a norm or convention to affect an agent's decision, it has to be recognized by the agent, that is, the agent has to believe that the norm or convention exists and that if she does not conform to it, she will incur a violation The latter is called *normative belief* by [32] (see also [6]). Recognition of a convention is guaranteed, if the agent belongs to the group of agents in which the convention holds. Indeed, as emphasized in Section 3.2, according to Lewis' definition, a certain regularity of behavior R is a convention for a population of agents P if and only if the agents in P mutually expect everyone in P to conform to the regularity R and commonly believe so. Thus, if agent i is a member of P and R is convention for P, then i has to believe that R is convention for P. The latter follows from the fact that if the agents in P have a common belief that some proposition p holds, then every agent in P has to believe so.[17]

Once the norm or convention with its associated costs and sanction for violation has be recognized by an agent, the agent will take it into consideration in her decision-making process. For the sake of clarity, we here distinguish *norm compliance* from mere *norm following*. Norm compliance requires the *goal* to conform to the content of the norm. In other words, for an agent to comply with a norm, she has to be motivated by the goal of conforming to what the norm prescribes. For example, an agent complies with the norm of paying taxes if she wants to pay taxes, after having recognized the corresponding norm that she ought to pay taxes. Norm following just requires that the agent chooses an action *knowing that* this choice will lead her to conform to what the norm prescribes. To sum up, while norm compliance requires *purposively* (or *intentionally*) conforming to what the norm prescribes, norm following only requires *knowingly* conforming to what the norm prescribes. Under the assumption that "purposively doing" implies "knowingly doing", norm compliance can be seen as a special case of norm following.

Two different forms of norm compliance exist. As we have emphasized in Section 2.1, some norms are internalized by the agent and give rise to moral values. If the agent decides

---

[17]This property can be formally proved in the logic of common belief [44].

to comply with them, she does it for ethical or moral reasons. In these cases, the agent's goal of conforming to what the norm prescribes is mainly originated from moral considerations. This is *ethical or moral compliance*. For example, an agent may comply with the legal obligation to pay taxes for ethical or moral reasons: the agent wants to pay taxes because she is motivated by the moral value to behave honestly. More generally, ethical compliance requires that the agent's goal of conforming to what the norm prescribes does not depend on the agent's actual desires[18] but only on the agents' actual moral values.[19]

In other cases, the agent complies with the norm because she desires to avoid the sanction or the social cost as a consequence of the violation and because she fears punishment. This is *opportunistic compliance* which is typical for conventions such as the following one:

> Except for pizza, sandwiches and other "finger foods", don't eat with your fingers.

This is a convention in Europe, as every person in Europe expects other people in Europe to follow it and every group of European people has a common belief that each of them expects the others to follow the convention. An European person believes that the convention exists and wants to follow it because she desires to avoid the social cost associated with the violation (e.g., the cost of being publicly blamed if she eats the food with her fingers).

In the case of opportunistic compliance, the agent wants to conform with what the norm prescribes because the consequences of norm violation (e.g., sanction, social cost, punishment) are undesirable for her, while the consequences of norm fulfillment (e.g., reward, social approval) are desirable for her. More generally, opportunistic compliance requires that the agent's goal of conforming to what the norm prescribes does not depend on the agent's actual moral values but only on the agents' actual desires.

We conclude the paper with the general observation that, although norm compliance has been extensively studied in the area of multi-agent systems, with an emphasis on both its logical aspects [3; 122; 74], and computational aspects [33; 4; 88; 149; 79], there is still no formal model which captures the distinctions between norm following and norm compliance, and between ethical compliance and opportunistic compliance. We believe this is an important issue. Its understanding would allow to complement a bottom-up approach to institutions, grounding them on the mental level via the collective level, with a top-down approach, explaining how institutions and norms influence the agents' cognition.

---

[18]This means that if the agent did have different desires in her mind, he would have had still the goal to follow the norm.

[19]This means that it is possible for the agent to reconsider her actual moral values in such a way that her goal to follow the norm is also reconsidered.

References

[1] C. Adam, A. Herzig, and D. Longin. A logical formalization of the OCC theory of emotions. *Synthese*, 168(2):201–248, 2009.

[2] T. Ågotnes, W. van der Hoek, and M. Wooldridge. Quantified coalition logic. In *Proceedings of the Twentieth International Joint Conference on Artificial Intelligence (IJCAI'07)*, pages 1181–1186. AAAI Press, 2007.

[3] T. Ågotnes, W. van der Hoek, and M. Wooldridge. Robust normative systems and a logic of norm compliance. *Logic Journal of the IGPL*, 18(1):4–30, 2009.

[4] N. Alechina, M. Dastani, and B. Logan. Programming norm-aware agents. In *Proceedings of the 11th International Conference on Autonomous Agents and Multiagent Systems (AAMAS 2012)*, pages 1057–1064. ACM Press, 2012.

[5] R. Alur, T. Henzinger, and O. Kupferman. Alternating-time temporal logic. *Journal of the ACM*, 49:672–713, 2002.

[6] G. Andrighetto, M. Campennì, F. Cecconi, and R. Conte. The complex loop of norm emergence: a simulation model. In K. Takadama, C. C. Revilla, and G. Deffuant, editors, *The Second World Congress on Social Simulation*, LNAI. Springer-Verlag, 2010.

[7] G. E. M. Anscombe. *Intention*. Basil Blackwell, 1957.

[8] R. Aumann. Interactive epistemology I: Knowledge. *International Journal of Game Theory*, 28(3):263–300, 1999.

[9] A. Baltag, S. Smets, and J. A. Zvesper. Keep ąőhopingąŕ for rationality: a solution to the backward induction paradox. *Synthese*, 169(2):301–333, 2009.

[10] P. Battigalli and M. Dufwenberg. Guilt in games. *The American Economic Review*, 97(2):170–176, 2007.

[11] P. Battigalli and M. Siniscalchi. Strong belief and forward induction reasoning. *J. of Economic Theory*, 106(2):356 391, 2002.

[12] R. F. Baumeister, A. M. Stillwell, and T. F. Heatherton. Guilt: an interpersonal approach. *Psychological Bullettin*, 115(2):243–267, 1994.

[13] N. Belnap, M. Perloff, and M. Xu. *Facing the future: agents and choices in our indeterminist world*. Oxford University Press, New York, 2001.

[14] C. Bicchieri. *The grammar of society: the nature and dynamics of social norms*. Cambridge University Press, 2006.

[15] K. Binmore. *Fun and Games: A Text on Game Theory*. D. C. Heath and Company, 1991.

[16] K. Binmore. *Natural Justice*. Oxford University Press, 2005.

[17] A. Bjorndahl, J. Y. Halpern, and R. Pass. Axiomatizing rationality. In *Proceedings of the Fourteenth International Conference on Principles of Knowledge Representation and Reasoning: (KR 2014)*. AAAI Press, 2014.

[18] C. Boutilier. Towards a logic for qualitative decision theory. In *Proceedings of International Conference on Principles of Knowledge Representation and Reasoning (KR'94)*, pages 75–86. AAAI Press, 1994.

[19] R. I. Brafman and Moshe Tennenholtz. An axiomatic treatment of three qualitative decision

criteria. *Journal of the ACM*, 47(3):452–482.

[20] R. I. Brafman and Moshe Tennenholtz. On the foundations of qualitative decision theory. In *Proceedings of the Thirteenth National Conference on Artificial Intelligence (AAAI'96)*, pages 1291–1296. AAAI Press, 1996.

[21] A. Brandenburger, A. Friedenberg, and J. Keisler. Admissibility in games. *Econometrica*, 76:307–352, 2008.

[22] M. Bratman. *Intentions, plans, and practical reason*. Harvard University Press, Cambridge, 1987.

[23] M. Bratman, D. J. Israel, and M. E. Pollack. Plans and resource-bounded practical reasoning. *Computational Intelligence*, 4:349–355, 1988.

[24] M. Bratman. Shared cooperative activity. *The Philosophical Review*, 101(2):327–41, 1992.

[25] J. Broersen. Deontic epistemic stit logic distinguishing modes of mens rea. *Journal of Applied Logic*, 9(2):137–152, 2011.

[26] J. Broersen, M. Dastani, J. Hulstijn, and L. van der Torre. Goal generation in the boid architecture. *Cognitive Science Quarterly*, 2(3-4):428–447, 2002.

[27] A. Caplin and J Leahy. Psycological expected utility theory and anticipatory feelings. *Quarterly Journal of Economics*, 116(1):55–79, 2001.

[28] Cristiano Castelfranchi. Modelling social action for AI agents. *Artificial Intelligence*, 103:157–182, 1998.

[29] G. Charness and M. Dufwenberg. Guilt in games. *Econometrica*, 74(6):1579–1601, 2009.

[30] H. Clark and C. Marshall. Definite reference and mutual knowledge. In A. K. Joshi, B. L. Webber, and I. A. Sag, editors, *Elements of discourse understanding*. 1981.

[31] P. R. Cohen and H. J. Levesque. Reasons: Belief support and goal dynamics. *Artificial Intelligence*, 42:213–61, 1990.

[32] R. Conte and C. Castelfranchi. From conventions to prescriptions. towards an integrated view of norms. *Artificial Intelligence and Law*, 7:323–340, 1999.

[33] N. Criado Pacheco, E. Argente, P. Noriega, and V. Botti. Human-inspired model for norm compliance decision making. *Information Sciences*, 245:218–239, 2013.

[34] R. P. Cubitt and R. Sugden. Common knowledge, salience and convention: a reconstruction of david lewis' game theory. *Economics and Philosophy*, 19:175–210, 2003.

[35] A. Damasio. *Descartes Error: Emotion, Reason and the Human Brain*. Putnam Publishing, New York, 1994.

[36] M. Dastani and E. Lorini. A logic of emotions: from appraisal to coping.

[37] D. Davidson. Intending. In *Essays on Actions and Events*. Oxford University Press, New York, 1980.

[38] D. C. Dennett. *The Intentional Stance*. MIT Press, Cambridge, Massachusetts, 1987.

[39] F. Dretske. *Explaining behavior: reasons in a world of causes*. MIT Press, 1988.

[40] B. Dunin-Keplicz and R. Verbrugge. Collective intentions. *Fundamenta Informaticae*, 51(3):271–295, 2002.

[41] B. Dunin-Keplicz and R. Verbrugge. *Teamwork in Multi-Agent Systems: A Formal Approach*.

Wiley, 2010.

[42] E. Durkheim. *The rules of Sociological Method*. Free Press, New York, 1982. first published in French in 1895.

[43] M. Esteva, J. Padget, and C. Sierra. Formalizing a language for institutions and norms. In *Intelligent Agents VIII (ATAL'01)*, volume 2333 of *LNAI*, pages 348–366, Berlin, 2001. Springer Verlag.

[44] R. Fagin, J. Halpern, Y. Moses, and M. Vardi. *Reasoning about Knowledge*. MIT Press, Cambridge, 1995.

[45] N. H. Frijda, P. Kuipers, and E. Ter Schure. Relations among emotion, appraisal, and emotional action readiness. *Journal of Personality and Social Psychology*, 57(2):212–228, 1989.

[46] P. Galeazzi and E. Lorini. Epistemic logic meets epistemic game theory: a comparison between multi-agent Kripke models and type spaces. *Synthese*, forthcoming, 2016.

[47] M. Gilbert. Modelling collective belief. *Synthese*, 73(1):185–204, 1987.

[48] M. Gilbert. *On Social Facts*. Routledge, London and New York, 1989.

[49] M. Goldszmidt and J. Pearl. Qualitative probability for default reasoning, belief revision and causal modeling. *Artificial Intelligence*, 84:52–112, 1996.

[50] R. M. Gordon. *The structure of emotions*. Cambridge University Press, Cambridge, 1987.

[51] E. Gradel. Model checking games. *Electronic Notes in Theoretical Computer Science*, 67:15–34, 2002.

[52] J. Gratch and S. Marsella. A domain independent framework for modeling emotion. *Journal of Cognitive Systems Research*, 5(4):269–306, 2004.

[53] D. Grossi, J.-J. Ch. Meyer, and F. Dignum. Classificatory aspects of counts-as: An analysis in modal logic. *Journal of Logic and Computation*, 16(5):613–643, 2006.

[54] D. Grossi and G. Pigozzi. *Judgment Aggregation: A Primer*. Synthesis Lectures on Artificial Intelligence and Machine Learning. Morgan & Claypool Publishers, 2014.

[55] Barbara Grosz and Sarit Kraus. Collaborative plans for complex group action. *Artificial Intelligence*, 86(2):269–357, 1996.

[56] J. Haidt. The moral emotions. In R. J. Davidson, K. R. Scherer, and H. H. Goldsmith, editors, *Handbook of affective sciences*, pages 852–870. 2003.

[57] A. Hájek and P. Pettit. A theory of human motivation. *Australian Journal of Philosophy*, 82:77–92, 2004.

[58] J. Y. Halpern. Beyond nash equilibrium: Solution concepts for the 21st century. In K. R. Apt and E. Gradel, editors, *Lectures in Game Theory for Computer Scientists*, pages 264–289. 2011.

[59] J. Y. Halpern and R. Pass. A logical characterization of iterated admissibility. In A. Heifetz, editor, *Proc. of TARK 2009*, pages 146–155, 2009.

[60] J. Harsanyi. Morality and the theory of rational behaviour. In A.K. Sen and B. Williams, editors, *Utilitarianism and Beyond*. Cambridge University Press, Cambridge, 1982.

[61] J. C. Harsanyi. Games with incomplete information played by 'bayesian' players. *Management Science*, 14:159–182, 1967.

[62] H. L. A. Hart. *The concept of law*. Clarendon Press, Oxford, 1992. new edition.

[63] A. Herzig and D. Longin. C&L intention revisited. In *Proceedings of the 9th International Conference on Principles on Principles of Knowledge Representation and Reasoning (KR 2004)*, pages 527–535. AAAI Press, 2004.

[64] J. Hintikka and G. Sandu. Game-theoretical semantics. In J. van Benthem and A. ter Meulen, editors, *Handbook of Logic and Language*, pages 361–410. Elsevier, 1997.

[65] A. Hopfensitz and E. Reuben. The importance of emotions for the effectiveness of social punishment. *The Economic Journal*, 119(540):1534–1559, 2009.

[66] J. F. Horty. *Agency and Deontic Logic*. Oxford University Press, Oxford, 2001.

[67] I. L. Humberstone. Direction of fit. *Mind*, 101(401):59–83, 1992.

[68] J. Doyle and R. Thomason. Background to qualitative decision theory. *The AI Magazine*, 20(2):55–68, 1999.

[69] A. Jones and M. J. Sergot. A formal characterization institutionalised power. *Journal of the IGPL*, 4:429–445, 1996.

[70] D. Kahneman and D. T. Miller. Norm theory: comparing reality to its alternatives. *Psychological Review*, 93(2):136–153, 1986.

[71] M. Kandori, G. Mailath, and R. Rob. Learning, mutation, and long run equilibria in games. *Econometrica*, 61:29–56, 1993.

[72] L. Keiff. Dialogical logic. In E. N. Zalta, editor, *The Stanford Encyclopedia of Philosophy*. 2011.

[73] D. Klein and E. Pacuit. Changing types: Information dynamics for qualitative type spaces. *Studia Logica*, 102:297–319, 2014.

[74] M. Knobbout and M. Dastani. Reasoning under compliance assumptions in normative multi-agent systems. In *Proceedings of the 11th International Conference on Autonomous Agents and Multiagent Systems (AAMAS 2012)*, pages 331–340. ACM Press, 2012.

[75] K. Konolige and M. E. Pollack. A representationalist theory of intention. In R. Bajcsy, editor, *Proceedings 13th International Joint Conference on Artificial Intelligence (IJCAI 93)*, pages 390–395, San Francisco, CA, 1993. Morgan Kaufmann Publishers.

[76] S. Kraus and D. J. Lehmann. Knowledge, belief and time. *Theoretical Computer Science*, 58:155–174, 1988.

[77] R. S. Lazarus. *Emotion and adaptation*. Oxford University Press, New York, 1991.

[78] J. LeDoux. *The emotional Brain*. Simon and Schuster, New York, 1996.

[79] J. Lee, J. Padget, B. Logan, D. Dybalova, and N. Alechina. Run-time norm compliance in BDI agents. In *Proceedings of the Proceedings of the 2014 international conference on Autonomous agents and multi-agent systems (AAMAS 2014)*, pages 1581–1582. ACM Press, 2014.

[80] D. Lewis. Desire as belief. *Mind*, 97:323–332, 1988.

[81] D. Lewis. Desire as belief ii. *Mind*, 105:303–313, 1996.

[82] D. K. Lewis. *Convention: a philosophical study*. Harvard University Press, Cambridge, 1969.

[83] C. List. The theory of judgment aggregation: an introductory review. *Synthese*, 187(1):179–

207, 2012.

[84] C. List. Three kinds of collective attitudes. *Erkenntnis*, 79(9):1601–1622, 2014.

[85] C. List and P. Pettit. *Group Agency: The Possibility, Design, and Status of Corporate Agents*. Oxford University Press, 2011.

[86] J. Locke. An essay concerning human understanding. Clarendon Press, Oxford, 1989.

[87] G. Loomes and R. Sugden. Testing for regret and disappointment in choice under uncertainty. *Economic J.*, 97:118–129, 1987.

[88] F. Lopez y Lopez, M. Luck, and M. d'Inverno. Normative agent reasoning in dynamic societies. In *Proceedings of the Third International Conference on Autonomous Agents and Multi-Agent Systems (AAMAS'04)*, pages 732–739. ACM Press, 2004.

[89] E. Lorini. Temporal STIT logic and its application to normative reasoning. *Journal of Applied Non-Classical Logics*, 23(4):372–399, 2013.

[90] E. Lorini. A logic for reasoning about moral agents. *Logique et Analyse*, 58(230):177–218, 2016.

[91] E. Lorini. A minimal logic for interactive epistemology. *Synthese*, 193(3):725–755, 2016.

[92] E. Lorini and C. Castelfranchi. The cognitive structure of Surprise: looking for basic principles. *Topoi: An International Review of Philosophy*, 26((1)):133–149, 2007.

[93] E. Lorini and A. Herzig. A logic of intention and attempt. *Synthese*, 163(1):45–77, 2008.

[94] E. Lorini and D. Longin. A logical account of institutions: from acceptances to norms via legislators. In *Proceedings of the International Conference on Principles of Knowledge Representation and Reasoning (KR 2008)*, pages 38–48. AAAI Press, 2008.

[95] E. Lorini, D. Longin, B. Gaudou, and A. Herzig. The logic of acceptance: grounding institutions on agents' attitudes. *Journal of Logic and Computation*, 19(6):901–940, 2009.

[96] E. Lorini, D. Longin, and E. Mayor. A logical analysis of responsibility attribution : emotions, individuals and collectives. *Journal of Logic and Computation*, 24(6):1313–1339, 2014.

[97] E. Lorini and R. Muehlenbernd. The long-term benefits of following fairness norms: a game-theoretic analysis. In *Proceedings of the 18th Conference on Principles and Practice of Multi-Agent Systems (PRIMA 2015)*, pages 301–318, Berlin, 2015. Springer-Verlag.

[98] E. Lorini and F. Schwarzentruber. A Modal Logic of Epistemic Games. *Games, Epistemic Game Theory and Modal Logic*, 1(4):478–526, 2010.

[99] E. Lorini and F. Schwarzentruber. A logic for reasoning about counterfactual emotions. *Artificial Intelligence*, 175(3-4):814–847, 2011.

[100] K. Ludwig and M. Jankovic. Collective intentionality. In L. McIntyre and A. Rosenberg, editors, *The Routledge Companion to the Philosophy of Social Science*. Routledge, New York, 2016.

[101] C. Mantzavinos, D.C. North, and S. Shariq. Learning, institutions, and economic performance. *Perspectives on Politics*, 2:75–84, 2004.

[102] J. G. March and H. A. Simon. *Organizations*. Wiley, New York, 1958.

[103] A. H. Maslow. A theory of human motivation. *Psychological Review*, 50:370–396, 1943.

[104] H. McCann. Settled objectives and rational constraints. *American Philosophical Quarterly*,

28:25–36, 1991.

[105] A. R. Mele. *Springs of Action: Understanding Intentional Behavior*. Oxford University Press, Oxford, 1992.

[106] J.-J. Ch. Meyer. Reasoning about emotional agents. *International J. of Intelligent Systems*, 21(6):601–619, 2006.

[107] J. J. Ch. Meyer, W. van der Hoek, and B. van Linder. A logical approach to the dynamics of commitments. *Artificial Intelligence*, 113(1-2):1–40, 1999.

[108] K. Miller and G. Sandu. Weak commitments. In G. Holmstron-Hintikka and R. Tuomela, editors, *Contemporary Action Theory, vol.2: Social Action*. Kluwer Academic Publishers, Dordrecht, 1997.

[109] R. Myerson. *Game Theory: Analysis of Conflict*. Harvard University Press, 1991.

[110] D.C. North. *Institutions, Institutional Change, and Economic Performance*. Cambridge University Press, Cambridge, 1990.

[111] A. Ortony, G.L. Clore, and A. Collins. *The cognitive structure of emotions*. Cambridge University Press, Cambridge, MA, 1988.

[112] M. Pauly. A modal logic for coalitional power in games. *Journal of Logic and Computation*, 12(1):149–166, 2002.

[113] A. Perea. *Epistemic game theory: reasoning and choice*. Cambridge University Press, 2012.

[114] M. Platts. *Ways of meaning*. Routledge and Kegan Paul, 1979.

[115] J. A. Plaza. Logics of public communications. In M. Emrich, M. Pfeifer, M. Hadzikadic, and Z. Ras, editors, *Proceedings of the 4th International Symposium on Methodologies for Intelligent Systems*, 201-216, 1989.

[116] Z. Pylyshyn. *Computation and Cognition: Toward a Foundation for Cognitive Science*. MIT Press, Cambridge, Massachusetts, 1984.

[117] A. S. Rao and M. P. Georgeff. Modelling rational agents within a BDI architecture. In *Proceedings of KR'91*, San Francisco, CA, 1991. Morgan Kaufmann Publishers.

[118] R. Reisenzein. Emotional experience in the computational belief-desire theory of emotion. *Emotion Review*, 1(3):214–222, 2009.

[119] N. Rescher. Semantic foundations for the logic of preference. In N. Rescher, editor, *The logic of decision and action*. University of Pittsburgh Press, 1967.

[120] S. Rick and G. Loewenstein. The role of emotion in economic behavior. In M. Lewis, J. Haviland-Jones, and L. Feldman-Barrett, editors, *The Handbook of Emotion*. Guilford, New York, 2008.

[121] I.J. Roseman, A.A. Antoniou, and P.E. Jose. Appraisal determinants of emotions: Constructing a more accurate and comprehensive theory. *Cognition and Emotion*, 10:241–277, 1996.

[122] A. Rotolo. Norm compliance of rule-based cognitive agents. In *Proceedings of the 22nd International Joint Conference on Artificial Intelligence (IJCAI 2011)*, pages 2716–2721. AAAI Press, 2011.

[123] K. R. Scherer, A. Schorr, and T. Johnstone, editors. *Appraisal Processes in Emotion: Theory, Methods, Research*. Oxford University Press, Oxford, 2001.

[124] T. Schroeder. *Three faces of desires*. Oxford University Press, 2004.

[125] J. Searle. *Expression and meaning*. Cambridge University Press, 1979.

[126] J. Searle. *Intentionality: An Essay in the Philosophy of Mind*. Cambridge University Press, New York, 1983.

[127] J. Searle. *Rationality in Action*. MIT Press, Cambridge, 2001.

[128] J. Searle. *The Construction of Social Reality*. The Free Press, New York, 1995.

[129] S. Sen and S. Airiau. Emergence of norms through social learning. In *Proceedings of the 20th International Joint Conference on Artificial Intelligence (IJCAI 2007)*, pages 1507–1512. ACM Press, 2007.

[130] Y. Shoham. Agent-oriented programming. *Artificial Intelligence*, 60:51–92, 1993.

[131] Y. Shoham and M. Tennenholtz. On the emergence of social conventions: modeling, analysis, and simulations. *Artificial Intelligence*, 94(1-2):139–166, 1997.

[132] G. Sillari. A logical framework for convention. *Synthese*, 147(2):379–400, 2005.

[133] H. A. Simon. Rational choice and the structure of the environment. *Psychological Review*, 63(2):129–138, 1956.

[134] M. Singh and N. Asher. A logic of intentions and beliefs. *Journal of Philosophical Logic*, 22:513–544, 1993.

[135] R. Stalnaker. Belief revision in games: forward and backward induction. *Mathematical Social Sciences*, 36:31–56, 1998.

[136] R. Stalnaker. Common ground. *Linguistics and Philosophy*, 25(5-6):701–721, 2002.

[137] R. Stalnaker. On logics of knowledge and belief. *Philosophical Studies*, 128:169–199, 2006.

[138] R. Sugden. *Economics of rights, co-operation and welfare (2nd Edition)*. Palgrave Macmillan, 2004.

[139] J. P. Tangney. Recent advances in the empirical study of shame and guilt. *American Behavioral Scientist*, 38(8):1132–1145, 1995.

[140] D. P. Tollefsen. Collective intentionality and the social sciences. *Philosophy of the Social Sciences*, 32(1):25–50, 2002.

[141] L. Tummolini, G. Andrighetto, C. Castelfranchi, and R. Conte. A convention or (tacit) agreement betwixt us: on reliance and its normative consequences. *Synthese*, 190(4):585–618, 2013.

[142] R. Tuomela. *The Philosophy of Social Practices: A Collective Acceptance View*. Cambridge University Press, Cambridge, 2002.

[143] R. Tuomela. *The Philosophy of Sociality*. Oxford University Press, Oxford, 2007.

[144] P. Turrini, J.-J. Ch. Meyer, and C. Castelfranchi. Coping with shame and sense of guilt: a Dynamic Logic Account. *Journal Autonomous Agents and Multi-Agent Systems*, 20(3):401–420, 2010.

[145] J. van Benthem. Rational dynamics and epistemic logic in games. *International Game Theory Review*, 9(1):13–45, 2007.

[146] J. van Benthem, P. Girard, and O. Roy. Everything else being equal: A modal logic for ceteris paribus preferences. *Journal of Philosophical Logic*, 38:83–125, 2009.

[147] H. P. van Ditmarsch, W. van der Hoek, and B. Kooi. *Dynamic Epistemic Logic*. Kluwer Academic Publishers, 2007.

[148] B. van Linder, van der Hoek, and J.-J. Ch. W., Meyer. Formalising abilities and opportunities. *Fundamenta Informaticae*, 34:53–101, 1998.

[149] M. B. van Riemsdijk, L. A. Dennis, M. Fisher, and K. V. Hindriks. Agent reasoning for norm compliance: a semantic approach. In *Proceedings of the 2013 international conference on Autonomous agents and multi-agent systems (AAMAS 2013)*, pages 499–506. ACM Press, 2013.

[150] P. Vanderschraaf. Convention as correlated equilibrium. *Erkenntnis*, 42(1):65–87, 1995.

[151] D. Villatoro, S. Sen, and J. Sabater-Mir. Exploring the dimensions of convention emergence in multiagent systems. *Advances in Complex Systems*, 14(2):201–227, 2011.

[152] G. H. Von Wright. *The logic of preference*. Edinburgh University Press, 1963.

[153] W. Walker and M. Wooldridge. Understanding the emergence of conventions in multi-agent systems. In *Proceedings of the First International Conference on Multi-Agent Systems (ICMAS-95)*, pages 384–389. AAAI Press, 1995.

[154] M. P. Wellman and J. Doyle. Preferential semantics for goals. In *Proceedings of the Ninth National conference on Artificial intelligence (AAAI'91)*, pages 698–703. AAAI Press, 2001.

[155] M. Wooldridge. *Reasoning about Rational Agents*. MIT Press, Cambridge, 2000.

[156] H. P. Young. The evolution of conventions. *Econometrica*, 61:57–84, 1993.

[157] M. Zeelenberg, W. van Dijk, A. S. R. Manstead, and J. van der Pligt. On bad decisions and disconfirmed expectancies: the psychology of regret and disappointement. *Cognition and Emotion*, 14(4):521–541, 2000.

[158] J. A. Zvesper. *Playing with Information*. PhD thesis, University of Amsterdam, The Netherlands, 2010.

www.ingramcontent.com/pod-product-compliance
Lightning Source LLC
Chambersburg PA
CBHW081129170426
43197CB00017B/2803